Intelligent Data-Analytics for Condition Monitoring

Intelligent Data-Analytics for Condition Monitoring

Smart Grid Applications

Hasmat Malik
BEARS, University Town, NUS Campus, Singapore;
Division of Instrumentation and Control Engineering,
Netaji Subhas Institute of Technology, Delhi, India

Nuzhat Fatema
Intelligent Prognostic Private Limited India;
Faculty of Business and Management, UniSZA, Malaysia

Atif Iqbal
Department of Electrical Engineering, Qatar University,
Doha, Qatar

ELSEVIER

ACADEMIC PRESS
An imprint of Elsevier

Academic Press is an imprint of Elsevier
125 London Wall, London EC2Y 5AS, United Kingdom
525 B Street, Suite 1650, San Diego, CA 92101, United States
50 Hampshire Street, 5th Floor, Cambridge, MA 02139, United States
The Boulevard, Langford Lane, Kidlington, Oxford OX5 1GB, United Kingdom

Library of Congress Cataloging-in-Publication Data
A catalog record for this book is available from the Library of Congress

British Library Cataloguing-in-Publication Data
A catalogue record for this book is available from the British Library

ISBN: 978-0-323-85510-5

For information on all Academic Press publications
visit our website at https://www.elsevier.com/books-and-journals

Publisher: Joe Hayton
Acquisitions Editor: Lisa Reading
Editorial Project Manager: Chris Hockaday
Production Project Manager: Nirmala Arumugam
Designer: Miles Hitchen

Typeset by Thomson Digital

Contents

Editors Biography

Dr. Hasmat Malik
BEARS, University Town, NUS Campus, Singapore; Division of Instrumentation and Control Engineering, Netaji Subhas Institute of Technology, Delhi, India

Hasmat Malik (SM'20) received BTech degree in electrical and electronics engineering from the GGSIP University, Delhi, India, MTech degree in electrical engineering from National Institute of Technology (NIT) Hamirpur, Himachal Pradesh, India, and the PhD degree in Electrical Engineering from Indian Institute of Technology (IIT), Delhi. He is a Chartered Engineer and Professional Engineer.

He is currently a Postdoctoral Fellow at BEARS, University-Town, NUS Campus, Singapore since January, 2019 and served as an Assistant Professor for 5+ years at Division of Instrumentation and Control Engineering, Netaji Subhas University of Technology (NSUT) Delhi, India. He has organized five international conferences and his proceedings have been published by Springer Nature. He is a Senior Member of the Institute of Electrical and Electronics Engineers (IEEE), USA, Life Member of ISTE (Indian Society for Technical Education), IETE (Institution of Electronics and Telecommunication Engineering), IAENG (International Association of Engineers, Hong Kong), ISRD (International Society for Research and Development) London, and CSTA (Computer Science Teachers Association) USA, Association for Computing Machinery (ACM) EIG, and Mir Labs, Asia.

He has published his research findings related to Intelligent Data Analytic, Artificial Intelligence, and Machine Learning applications in Power system, Power apparatus, Smart building & automation, Smart grid, Forecasting, Prediction and Renewable Energy Sources widely in International Journals and Conferences. Dr. Hasmat has authored/co-authored more than 100 research papers, 8 books, and 13 chapters in 9 other books, published by IEEE, Springer, and Elsevier.

He is a Guest Editor of Special Issue of Journal of Intelligent & Fuzzy Systems, 2018, 2020 (SCI, Impact Factor 2019:1.85), (IOS Press). He received the POSOCO Power System Award (PPSA-2017) for his PhD work for research and innovation in the area of power system. He has received best research papers awards at IEEE INDICON-2015, and full registration fee award at IEEE SSD-2012 (Germany). He has supervised 23 PG students. He is involved in several large R&D projects. His principle area of research interests is artificial intelligence, machine learning, and big-data analytics for renewable energy, smart building and automation, condition monitoring, and online fault detection and diagnosis (FDD).

Dr. Nuzhat Fatema
Intelligent Prognostic Private Limited India; Faculty of Business and Management, UniSZA, Malaysia

Dr. Nuzhat Fatema is graduated from Maharashtra University of Health Sciences, India. She has cured many patients with her skills of medicinal knowledge. Later to go beyond the clinical skills, she has achieved post-graduation in hospital management from International Institute of Health Management Research (IIHMR), Delhi. This was the platform where she has utilized her clinical skills with her managerial skills using artificial intelligence (AI), Machine Learning (ML), and Data Analytics. She has worked as a research associate at National Board of Examinations (NBE) India. She has authored one book describing a trouble free tool prepared by using different standardized manuals of medicines in different countries for usage of most complicated drug like Warfarin. She has published several research papers in renowned international journals and conferences. Presently, she is associated with Intelligent Prognostic Private Limited and Faculty of Business and Management, Universiti Sultan Zainal Abidin (UniSZA), Malaysia.

Her area of interest is AI, ML, and intelligent data analytics application in healthcare, monitoring, prediction, forecasting, detection, and diagnosis where she believes that it is a data driven world with stockpile of database in the industry which is to be used to extract value to make better informed, more accurate decisions in diagnosis, management, and better outcomes in industry care. Simply throwing the numbers by analyzing any data has zero value; therefore she has produced narratives using data for decision making. She has been doing research study by spotting patterns in data and setting up infrastructure in real-time industrial monitoring domain.

Prof. Atif Iqbal
Department of Electrical Engineering, Qatar University, Doha, Qatar

Atif Iqbal, *Fellow IET (UK), Fellow IE (India), and Senior Member IEEE,* Vice-Chair, IEEE Qatar section, DSc (Poland), PhD (UK)- Associate Editor, IEEE Trans. On Industrial Electronics, IEEE ACCESS, Editor-in-Chief, I'manager Journal of Electrical Engineering, Former Associate Editor IEEE Trans. On Industry Application, Former Guest Associate Editor IEEE Trans. On Power Electronics. Full Professor at the Department of Electrical Engineering, Qatar University and Former Full Professor at the Department of Electrical Engineering, Aligarh Muslim University (AMU), Aligarh, India. Recipient of Outstanding Faculty Merit Award academic year 2014–15 and Research excellence awards 2015 and 2019 at Qatar University, Doha, Qatar. He received his BSc (Gold Medal) and MSc Engineering (Power System and Drives) degrees in 1991 and 1996, respectively, from the Aligarh Muslim University (AMU), Aligarh, India and PhD in 2006 from Liverpool John Moores University, Liverpool, UK. He obtained DSc (Habilitation) from Gdansk University of Technology in Control, Informatics and Electrical Engineering in 2019. He has been employed as a Lecturer in the Department of Electrical Engineering, AMU, Aligarh since 1991 where he served as Full Professor until August 2016. He is recipient of Maulana Tufail Ahmad Gold Medal for standing first at BSc Engg. (Electrical) Exams in 1991 from AMU. He has received several best research papers awards, for example, at IEEE ICIT-2013, IET-SEISCON-2013, SIGMA 2018, IEEE CENCON 2019, IEEE ICIOT 2020, and Springer ICRP 2020. He has published his research findings related to Power Electronics, Variable Speed Drives and Renewable Energy Sources widely in International Journals and Conferences. Dr. Iqbal has authored/co-authored more than 440 research papers, 4 books, and several chapters in edited books. He has supervised several large R&D projects worth more than multi million USD. He has supervised and co-supervised several PhD students. His principal area of research interest is Smart Grid, Complex Energy Transition, Active Distribution Network, Electric Vehicles drivetrain, Sustainable Development and Energy Security, Distributed Energy Generation, and multiphase motor drive system.

Preface

Nowadays, huge amount of data in all domains have been generated. It is important to bring intelligent and meaningful use of this data as required for an optimized and efficient engineering process. Data analytics is the analysis of raw data, which can be used to draw conclusions and use in decision making. Data analytics techniques can divulge trends that can be effectively be utilized for intelligent decision making. Data analytics is important in businesses to improve their performance, reduce cost, analyze customer behavior and satisfaction, and increase their profitability. Similarly, data analytics is important in a smart grid. With the development in the ICT, an additional layer is integrated in the smart grid to collect data using smart sensors and smart meters and analyzing them using Big Data analytics.

This is the subject of this book where data analytics is employed in smart grid context in several areas. The characterizations of machine learning, data collection, and storage are first illustrated as a prelude to demonstrating the motivation and potential advantages of implementing advanced data analytics in smart grids. Smart grid is the power system network that smartly integrates the action of all connected users, for example, generators, load, consumers, and deliver sustainable, secure, and economic energy supply.

This book contains 10 different chapters that discusses the fundamentals of machine learning and data analytics in the area of smart grid and its application. Two major parts of the book are:

- **Part A:** Intelligent data analytics for classification in smart grid
- **Part B:** Intelligent data analytics for forecasting in smart grid

The first part of the book includes application of intelligent data analytics in solar PV fault diagnostics, transformer health monitoring and faults diagnostics, and induction motor faults. Data analytics for power quality analysis of power system and transmission line diagnostics are further elaborated and discussed. The second part of the book illustrates the forecasting issues using data analytics. Forecasting is highly important for smart grid operation to achieve sustainable and reliable operation. Global solar radiation forecasting is discussed in the book followed by the wind data forecasting. Load forecasting is important from demand side management, which is taken up in one chapter. Finally data analytics is used for battery charging/discharging forecasting. Several advanced artificial intelligent/machine learning approaches such as Deep Convolutional Neural Network (ConvNet/CNN), Fuzzy Reinforcement Learning (FRL), Modified Fuzzy Q Learning (MFQL), Gene Expression Programming (GEP),

Extreme-Learning Machine (ELM), Semi-supervised & Unsupervised ELM, Proximal Support Vector Machine (PSVM), Deep Learning Neural Network, Long Short-Term memory (LSTM) Network, and Deep Learning Neural Network (DLNN) are employed and elaborated deeply in the book chapters in the area of smart grid and condition monitoring.

PART A: Intelligent Data Analytics for Classification in Smart Grid

Chapter 1: Advances in Machine Learning and Data Analytics. In this chapter, detailed information of data analytics of smart grid application, data analytics for business, condition monitoring, data and its relation, data pre-processing, feature extraction, feature selection, and different application areas are presented along with a wide list of software, dataset's digital library.

Chapter 2: Intelligent Data Analytics for PV Fault diagnosis Using Deep Convolutional Neural Network (ConvNet/CNN). In this chapter, a deep neural network using ConvNet/CNN based algorithm has been proposed for automatic fault diagnosis of photovoltaic module (PVM) and localize the anomaly condition in an interconnected PV system. The proposed DNN model using ConvNet/CNN algorithm does online diagnosis of PV module. The obtained results during training and testing phase shows its outperformance for solar PV module failure analysis.

Chapter 3: Intelligent Data Analytics for Power Transformer Health Monitoring Using Modified Fuzzy Q Learning (MFQL). In the smart grid application, the turbine in a wind farm/WECS (wind energy conversion system) is connected with a power transformer (WTPT), which increases the generated output voltage of generator from a few hundred volts to a medium voltage level of distribution system. Therefore, to maintain the healthy condition of the smart grid, an intelligent data analytics approach for power transformer health monitoring using modified fuzzy Q learning (MFQL) is proposed. The proposed approach is able to provide the condition monitoring, fault detection, and diagnosis (FDD) information to the system operation in an efficient time interval so that the operator might execute corrective action and/or plan a PdM (predictive maintenance) for the equipment.

Chapter 4: Intelligent Data Analytics for Induction Motor Using Gene Expression Programming (GEP). In this chapter, a realistic FDCA method for external fault identification for three phase IMs using Gene Expression Programming (GEP) have been proposed and is validated on publicly available real fault data. The GEP approach uses RMS values of 3-phase voltages and currents as input variables for identifying six types of external faults experienced by IM and one normal operating (NF) condition. GEP approach is compared against artificial neural network (ANN) (i.e., multilayer perceptron neural network-MLP) and support vector machine (SVM) techniques, which reveals that GEP approach is superior in terms of analytic accuracy and has lower computational requirements.

Chapter 5: Intelligent Data Analytics for Power Quality Disturbance Analysis Using Multi-Class ELM. In this chapter, a novel approach for power quality disturbance diagnosis (PQDD) is proposed, which includes the model development, real-time data generation, data pre-processing, feature extraction, and feature selection. The EMD approach is developed for feature extraction and obtained IMFs are processed through the decision tree based machine learning approach of J48 algorithm, which is used for feature selection and obtained results are compared with conventional AI method of MLP-ANN. The results show the outperformance of the proposed approach.

Chapter 6: Intelligent Data Analytics for Transmission Line Fault Diagnosis Using EEMD Based Multiclass SVM and PSVM. Today online condition monitoring, fault detection & diagnosis (FDD) of a transmission line, is required to predict the actual operating condition (AOC) of the electrical power network (EPN). In this chapter, an alternative approach to predict the AOC during an online operating scenario has been formulated, which can identify the AOC from easily recorded parameters of the transmission line. For this, EEMD (ensemble empirical mode decomposition) based PSVM (proximal support vector machine), the approach has been implemented which has significantly high processing speed as compared to other AI methods. The obtained results show that the new proposed framework is effective in evaluating the AOC without the need to measure the other parameters except for current and voltage signals.

PART B: Intelligent Data Analytics for Forecasting in Smart Grid

Chapter 7: Intelligent Data Analytics for Global Solar Radiation Forecasting for Solar Power Production Using Deep Learning Neural Network (DLNN). This chapter proposes DLNN based intelligent data analytics forecasting approach for multi-step ahead (MSA) global solar radiation (GSR), in which per minute recorded data of 3 years (during 2015 to 2017) was first collected from meteorological department of India, then data were pre-processed including cleaning the data, missing value filling, and spikes removal. The MSA forecasting is achieved recursively by utilizing the first forecasted data as an input to generate the next forecasting data and the process is achieved up to level four. The results are compared with other method (artificial neural network-ANN), which represents that proposed DLNN based forecasting approach has excellent performance and can be used for online forecasting applications.

Chapter 8: Intelligent Data Analytics for Wind Speed Forecasting for Wind Power Production Using Long Short-Term memory (LSTM) Network. Renewable energy sources (RES) increasing day-by-day, which shared around 65% contribution of the total global power generation in 2018, wind energy source plays an important role. This chapter proposes a LSTM (Long Short-Term memory) and modified LSTM (mLSTM) network based intelligent data analytics forecasting method for multi-step ahead (MSA) wind speed (WS) forecasting. Both LSTM and mLSTM have been developed for the same data, which shows that proposed WS forecasting approach is outperformed and can be utilized for other locations as well.

Chapter 9: Intelligent Data Analytics for Time-Series Load Forecasting Using Fuzzy Reinforcement Learning (FRL). In this chapter, a short-term load predictor which is able to forecast the load for next 24 h is proposed and is tested by using real-time recorded historical data collected from GEFCom2012 and GEFCom2014 and simulated results shows accurate and highly satisfactory performance. In this chapter, four different case studies have been performed for 1 month-ahead, week-ahead, day-ahead, and hour-ahead load forecasting.

Chapter 10: Intelligent Data Analytics for Battery Charging/Discharging Forecasting Using Semi-supervised and Unsupervised Extreme Learning Machines. In this chapter, an alternative way to estimate the remaining useful life (RUL) during an online operating scenario has been formulated, which can extract the RUL from easy-recorded parameter of lithium-ion (Li-ion) battery (LIB). To form a linear relationship between formulated health indicator (HI) and actual battery capacity, a power transformation method is applied and then Pearson and Spearman rank correlation methods is used to validate its performance. The supervised extreme learning machine (SELM), semi-supervised ELM (SSELM), and unsupervised ELM (USELM) are applied to evaluate the RUL by using extracted HI and transformed HI. This is the first attempt to implement the SSELM and USELM for RUL estimation of LIBs to make the system online in the real-time domain. The obtained results show that the new proposed approach is effective in evaluating the battery health condition without the need to measure the R and capacity.

Due to the simplicity of the intelligent data analytics methods and flexibility, readers from any field of study can employ them for classification, forecasting, prediction, etc. The book shall serve as a viable source on how to design, adapt, and evaluate the algorithms, which would be beneficial for the readers interested in learning and developing data analytics machine algorithms. The book will find adaptability among researchers and practicing engineers alike. This single volume encompasses wide range of real-time smart grid applications with fundamental to advanced level information.

We are sincerely thankful to the *Intelligent Prognostic Private Limited India* to provide all type of technical and non-technical facilities, cooperation, and support in each stage to make this book in reality.

We wish to thank our colleagues and friends for their insight and helpful discussion during the production of this book. We would like to highlight the contribution of Prof. Haitham Abu-Rub, Texas A&M University at Qatar, Prof. Sukumar Mishra, IIT Delhi at India, Prof. Bhim Singh, IIT Delhi, India, Prof. B.K. Panigrahi, IIT Delhi, India, Prof. Fausto Pedro García Márquez, UCLM Spain, Prof. Majid Jamil, Jamia Millia Islamia, Delhi, India, Prof. D.P. Kothari, Director Research, Gaikwad Patil Group of Institutions, Nagpur, India, Dr. Rizwan, DTU India, Prof. Imtiaz Ashraf, Aligarh Muslim University, India, Prof. M.S. Jamil Asghar, Aligarh Muslim University, India, Prof. Salman Hameed, Aligarh Muslim University, India, Prof. A.H. Bhat, NIT Srinagar, India, Prof. Kouzou Abdellah, Djelfa University, Algeria, Prof. Jaroslaw Guzinski, Gdansk University of Technology, Poland, Prof. Akhtar Kalam, Victoria University of Technology, Australia, Prof. Mairaj Ud Din Mufti, NIT Srinagar,

India, Prof. YR Sood, NIT Hamirpur (HP), India, Prof. A.P. Mittal, NSUT Delhi, India, Prof. R.K Jarial, NIT Hamirpur (HP), India, Prof. Rajesh Kumar, GGSIPU, India, Prof. Anand Parey, IIT Indore, India, Dr. Jafar A. Alzubi, Al-Balqa Applied University, Jorden, Dr. Majed Alotaibi, King Saud University, Riyadh, Saudi Arabia, and Dr. Abdulaziz Almutairi, Majmaah University, Majmaah, Saudi Arabia.

We further would like to express our love and affection to our family members, Shadma (wife of Prof. Atif Iqbal), Abuzar, Abubaker (Sons of Prof. Atif Iqbal), Noorin (daughter of Prof. Atif Iqbal). We would like to express our gratitude to Zainub Fatema and Ayesha Fatema (daughters of Dr. Hasmat Malik and Dr. Nuzhat Fatema) for their intense feeling of deep affection.

Woodlands, Singapore/Delhi, India — Dr. Hasmat Malik
Woodlands, Singapore/UniSZA, Malaysia — Dr. Nuzhat Fatema
Doha, Qatar — Prof. Dr. Atif Iqbal

Part A

Intelligent Data Analytics for Classification in Smart Grid

Chapter 1

Advances in Machine Learning and Data Analytics

1 Introduction

In the modern power world, the smart grid implementations are driving utilities to make advance their power networks so that wide area communication between M2M (machine to machine) and M2H (machine to human) may become modernize. There are several sub-system level's equipments (i.e., PMU-phasor measurement unit, energy management system, automated metering mechanism, instrumentation & control, etc.), which communicate between each other or node to node. Due to this communication, a huge volume of data flow is created in a second or microsecond within the power network, which is not an easy task to handle such data volume in a simple or via a conventional way. This generate a technical challenge to the professionals because message timing, framing, and processing may become complex. Furthermore, without efficient data handling mechanism, many hours are consumed in attempting to receive a new application to do work. Moreover, new applications require high performance abilities and advance processing mechanism to take a fast and correction action as well as decision. Due to this, negative consequences can result if the process occurs on live networks.

To overcome such type of problem in real world as mentioned earlier, both machine learning and data analytics are becoming popular research area for industries as well as universities. This popularity is driven by the internet and its associated huge number of IoTs (Internet of Things) devices connected in the smart grid because they generate a huge amount of data every second and proper understanding of such big-data needs advanced machine learning and/or data analytics approaches. Moreover, several applications such as VPP (virtual power plant) management, DSM (demand side management), DER (distributed energy resources) management, forecasting, prediction, ESS (energy storage system) management, and EVs (electric vehicles) management have extensively utilized machine learning (ML) approaches, which lead it to enhance the popularity of advance machine learning and intelligent data analytics. Generally, both data analytics and ML are closely related to each other. And both techniques are used for harnessing the useful information about the system behavior without the model's physical parameters information. Both techniques are capable for data driven

Intelligent Data-Analytics for Condition Monitoring. http://dx.doi.org/10.1016/B978-0-323-85510-5.00001-6

health monitoring of smart grid application. Therefore, application and advancement of both ML as well as data mining approaches is increasing in extensive way.

1.1 Brief information of data base analysis

In the research domain, the data analytics is very important for the technology commercialization for laying the foundation in the real world. So, doing research, the data analytics is very useful for: (1) understanding the trends and finding the major events; (2) visibility of timing, risks, and rewards; and (3) context analysis for market size identification and customer demography identification. Therefore, there are three main categories of the data which moves from the macro level to micro level of the study and/or research for any field. These data are: (1) country level data; (2) industry level data; and (3) company level data. Any type of data informs two main points such as statistics and analysis. Therefore, country level data gives the information in form of statistics of demography, GDP, wages, etc., and information of analysis—economic analysis, social analysis, political and legal type of analysis. The industry level data gives the information of statistics (as industry size, brand, and market share, etc.) and analysis (as trend analysis, highlights, SWOT, etc.). At the end, company data is a like micro data, which gives the information about statistics (i.e., financial data, company size, etc.) and analysis (i.e., SWOT, news, third party reports, etc.) as well. The SWOT is explained as the internal and external information with respect to the positive as well as negative sign of the data or coming information from the data (such as strength, weakness, opportunities, and treats, etc.). Therefore, there is an important terminology in the domain of IDA that is called "sweet spot," which tells the exact position/level of the innovation performed by the inventor. This is the common point in between feasibility (i.e., technological and environmental), desirability (i.e., social, legal, and political), and viability (i.e., economics of industrial point of view) [1].

Therefore, for doing database IDA for any innovation, innovators need database. According to previous explanation, innovator needs country level, industry level, and company level database for the further analysis. In this regards, there are some collected information which is represented in Table 1.6 of Section 7.

1.2 Brief information of intelligent data analytics for business

Intelligent data/big data analytics (IDA/IBDA) is a process to harness the information of the system and/or its associated sub-systems by utilizing the data instead of physical parameters or system modeling information. Now days in the IoT world, everything is going toward the IBDA and hence, its market value in term of the business forecast, has become valuable and important. Generally, expert analyses the market of IDA/IBDA into two broad groups, which are: data visualization and discovery (DVD) and analysis at advanced level (AAL). The major players/shareholder of this domains come from healthcare industry, biotech-companies, insurance sector, bank sector, retail sector, government agencies, defense and its associated companies, etc. The IDA/IBDA provides the support to the organization to make real-time decision and also provide the real-time position of the company so that organization may enhance the business. According to the historical data, the IDA/IBDA market was around $8.54 billion in 2017 and expected to grow the CARG (compound annual growth rate) more than 29% (~40.65 billion dollars) by coming years [2].

In the DVD approach and formulation of the architecture, mainly data-integration, data transformation (and analysis), and visualization are the main components. Numerous types of data come from various sources (i.e., machine data: sensors, RFID, camera; operational data; files: Excel/CSV, networks, web/social/cloud data; application based data, etc.), which are integrated and data fusion process is performed. In the data transformation, integrated data is transformed and/or decoded for the further application point of view. In the visualization, analyzed data is represented in form of graph/patterns etc., so that technologist/manager/analytics get benefited. Similarly, the AAL architecture includes the data integration, data model building, model development, model optimization, and visualization. Full procedure depends upon the data sources as mentioned for DVD.

Generally, market analysis is performed in the form of life cycle, which depends on mainly two components such as market value and time interval. The market value is varied from starting zero (0) value to higher side. While time period shows the actual level of the growth of the technology such as at development stage, growth stage, maturity stage, and decline phase. The market value of the IDA/IBDA is varied as per the reason/location of the market. For example, in Europe and the United States, IDA/IBDA is at growth level, while IDA/IBDA in Africa and Latin America is under development stage only. Therefore, there are a lot of options in IDA/IBDA domain in the region of Africa, Latin America, Asia, Middle East, and some part of Europe (except western part). According to the expert opinion, market growth in Asia-Pacific region is very high (in growth rate) in comparison to the rest of the world.

The revenue of IDA/IBDA mainly depends only on few industries and services areas as follow: GFS, G&I, retail, telecommunications, healthcare & life science, business services, information & communication technologies, manufacturing, media entertainment, and others. The highest and lowest revenue share of IDA/IBDA are from BFS and media & entertainment, respectively. Apart from this, healthcare & life science industry is approximately at the middle in terms of revenue share. In the form of market share govern by the several solution providers, few of them are: SAS, IBM, QLIK, Splunk, TIBCO, Tableau, FICO, Plantir, SAP, and Alteryx, etc., in which, SAS, IBM, and QLIK are the main market players which govern around half of the world market share in this domain. Key note for the reader is that presented information in this section is collected based on open access information received through Google search. The authors are not responsible for any incorrect information supplied by Google search or any other mode of search.

1.3 Brief information of intelligent data analytics for smart grid

According to the International Energy Agency (IEA) report on smart grids analysis [3], the overall investment is decreasing while its technology has become smarter. The United States has been at the position with an increased investment of 12%. Grid investment in Europe region remained same in 2019 of USD 50 billion. In India, the investment goes down by (−20%) in 2019 even though strong infrastructure of the T&D (transmission and distribution) lines. While smart grid investment focuses on the hardware side of digital substation, smart metering, etc. Now days, the smart grid industry is increasing day-by-day and will reach up to USD 112.7 billion in next 5 years. In the "AWS + C3.ai" report on Application Development Time and Cost Saving [4], the global impact of AI on additional economic value of 1.2 trillion was in 2016 and forecasts 3.5–5.8 trillion impact from AI by 2020.

1.4 Brief information of intelligent data analytics for condition monitoring

The condition monitoring (CM) is a way or method of monitoring the health of the system, sub-system, and/or its associated equipments. CM can be performed by using any type of methods such as physics-based method, model-based method, and data-driven methods. The physics-based and model-based methods required system information, while data-driven method works on historical data captured with respect to the time varying interval. Generally, mechanical and/or electrical data generated with the help of sensor are used for CM of the machinery. There are some basic steps to perform the CM of a system/sub-system/its associated equipments, these are: (1) identify the critical assets with the system; (2) identify the parameters to be monitored and its sampling frequency rate; (3) identify the suitable CM solution/method; (4) demonstration/installation of the CM solution on site; and (5) integration of the CM system with the available maintenance solution in the system. There are some basic significant indications of the failure condition in the system/machinery. These indications are the vibration, noise, heat, smell, smoke, and finally breakdown condition. So, to prevent the system, doing CM in advanced level is not an easy task for the maintenance engineers. Generally, three types of maintenance scenarios (PdM: predictive maintenance, TBM: time based maintenance/planned maintenance, and RM: reactive maintenance) are performed for doing CM. Normally, the indication of failures (i.e., vibration, noise, heat, smell, smoke, and finally breakdown condition) come under RM, which is fixed after the system breaks. But system owner wants to fix failure condition before breaks. Under this situation, PdM and TBM play an important role so that cost of repair may reduce with respect to enhance the system performance.

Several types of CM techniques are utilized by the industry such as: (1) vibration based CM; (2) lubricating oil analysis based CM; (3) thermographic based CM; and (4) other methods (MCSA: motor current signature analysis).

According to the market survey, the key contributions of these methods of CM are as per their ranked values: vibration CM, lubricating oil analysis, thermography, MCSA. While vibration based CM covers the market share more than 65% but the cost of analysis is higher than MCSA. The service market of CM is very good since 2016 to till now with higher growth rate of CAGR (compound annual growth rate). According to the market revenue of Europe region, Germany is at the top followed by the United Kingdom, Italy, and France. In the Asia-Pacific region, China is at the top followed by the Japan and India. According to the current statistics, top service providers are GE Measurement & Control followed by SKF, Emerson, and Rockwell Automation.

The initiation of IDA/IBDA in the area of CM carries big market revenue of the business part. Since starting in 2000 and till date, IDA/IBDA is categorized into four phases. Phase-1 is in between 2000 and 2005, which is related to the asset management. Phase-2 belongs to the asset management & data reporting from 2005 to 2010. IDA/IBDA, and evaluation comes in phase-3 from 2010 to 2015 and finally phase-4 belongs to PdM or predictive analysis, which started in 2015 to till date. Moreover, in the future, prescriptive analytics may come in picture which would lead through digital twin technology. The prescriptive analysis mainly focuses on the future prediction of the equipment condition.

In the CM area following broad research area will come as energy monitoring, smart building monitoring, production/process monitoring, and environmental monitoring. In the energy monitoring, data coming from smart meters will be used for IDA/IBDA. In the area of smart building monitoring, data coming from air-conditioning system, boiler, sensors, etc., are used. The soil, water, air, and emission data monitoring comes under the environmental monitoring. Therefore, there are some key players which hold the key position in the market as per the method of CM. For example, in the area of MCSA, the key contributors are: Siemen AG (at top), ABB, Mitsubishi, and Bosch Rexroth. In the area of thermography-based CM, the key contributors are: FLIR systems Inc (at top), Fluke corporation, and Nippon Avionics Co. In the area of lubrication oil based CM, the key contributors are: Bureau Veritas, Intertek Group, SGS, ALS Global, Spectro Scientific, TestOil, Parker Hannifin, GE, GasTOPS, and Castrol. Finally, in the area of vibration-based CM, the key contributors are: GE, SKF, Emerson, Rockwell Automation, and Bruel & Kjaer Vibro.

2 Data and it's relation

According to the Raymond Wolfinger, the plural of anecdote is data. In the modern world, data are everywhere and they can be in various forms (i.e., text, word, picture, video, character, signatures, etc.). Each type of data will be labelled or unlabelled in form of structured or unstructured data. In the CM of smart grid applications, different types of datasets are utilized for the further study as follows: electrical signature type of data (i.e., power, current, voltage, frequency, etc.), mechanical signature type of data (i.e., vibration, refrigerant, temperature, humidity, flow rare, pressure, heat rejection rate, cooling rate, heat transfer rate, coefficient of performance, efficiency, etc.), renewable energy related data (i.e., global horizontal irradiance, direct normal irradiance, diffused horizontal irradiance, wind speed, wind direction, maximum wind speed, air temperature, relative humidity, pressure, precipitation, dew point, wet bulb temperature, latitude, longitude, altitude, height from sea level, etc.), energy storage device related dataset (i.e., charging/discharging cycle count, voltage, current, internal resistance, Ah-capacity, impedance, temperature, time, etc.), power transformer related data (dissolved gases, Furan analysis, dielectric strength, metal particle count, moisture level, power factor dissipation factor, interfacial tension, acid number, furans, oxygen inhibitor, swift frequency response, DC resistance, dielectric loss, infrared camera based data, oil level, ground resistance, oil leakage, etc.), and other type of data are web and social media, machine to machine, transaction, biometric, human generated data, etc.

The term Big-Data is referred to the volume, variety, and velocity of the data. This term was described in 2001 by Laney, which is known as 3Vs of data. Furthermore, 10 Vs model of the dataset come in the market, which referred vagueness (related to confusion over meaning of big data and tools), variability (related to dynamic evolving behavior in the data source), venue (related to distributed heterogeneous data from multiple platform), variety (related to different types of data), velocity (related to speed at which data is generated), vocabulary (related to data model semantic that describe data structure), validity (related to data quality, governance, master data management), veracity (related to data accuracy), and volume (related to size of data).

The IoTs are rapidly increasing to make things smart and interconnect, hence sensors and volumes of data create a massive level of generated data volume as shown in Fig. 1.1. This data need to transfer between devices, services and useful application scenarios. In Industry 4.0, the connected IoT devices will

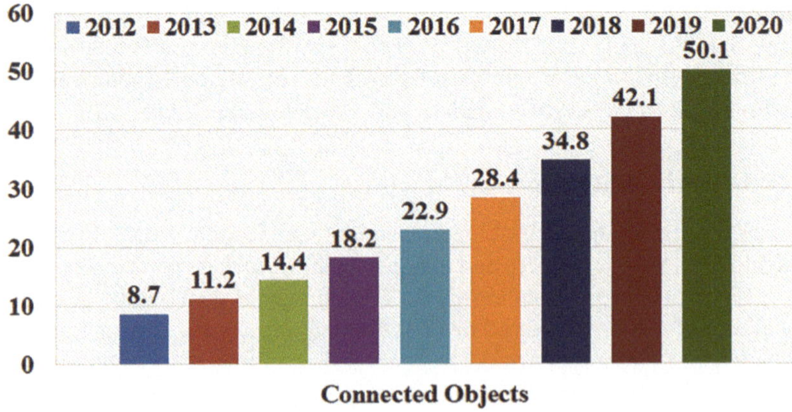

FIGURE 1.1 **Growth in IoT devices.** *(NCTA data.)*

be ~50.1 billion by 2020 [5] and it is expected to generate around 79.4 ZB of data by 2025 (as per IDC forecast).

3 Data preprocessing (DPP)

The data preprocessing (DPP) is an IDA technique that involves transforming the raw captured data into an understandable and useable format to solve the problems in the real-world application implementation. The real site recorded data is generally incomplete, inconsistent, and/or far away from trends and pattern due to some errors (i.e., missing values, spikes, non-linearity of the data, signal-to-noise ration, etc.). Therefore, data processing is a proven approach to resolve such type of problems and enhance the correlation of the data with respect to the labelled target value.

Hence, in the DPP approaches, three main methodologies are followed, which are as follow: (1) data processing; (2) feature extraction; and (3) feature selection. In the data processing approach (DPA), several issues related to data quality and its behavior, are resolved such as data cleaning, data transformation, data reduction, data integration, and finally data discretization. In the process of data cleaning, data is processed for removing the spikes and filling the missing values (if any). Then data transformation is performed to make the uniformity of the data for all case scenario of the machinery with respect to the perturb loading condition. Thereafter, data reduction is performed to rearrange the similar data in a one group. Then data integration is performed by combining the multiple dataset. Moreover, different IA/machine learning-based methods for the DP are: single user programming, multiple programming, real-time process, on-line processing, time sharing processing, and distributed processing.

After the DP, feature extraction and feature selection procedure is performed, which are explained in sub-sequence section as given later.

3.1 Feature extraction

The feature extraction process (FEP) includes reducing the number of input variables/attributes/resources to describe the similar information of a large volume of dataset [6–9]. For any kind of application (i.e., image processing, machine learning, pattern recognition, classification, regression, clustering, etc.), FEP starts from an initial step of data recording. Generally, FEP belongs to the dimensionality reduction process [6–12].

In the CM of smart grid application domain, feature extraction is performed by three main methodologies. These methodologies are: (1) time-domain based feature extraction (TDbFE); (2) frequency-domain based feature extraction (FDbFE); and (3) time-frequency-domain based feature extraction (TFDbFE). The represented Tables 1.1–1.3 show a broad list of features of TDbFE, FDbFE, and TFDbFE, respectively.

TABLE 1.1 Brief summary of TDbFE for CM of smart grid applications.

S. no.	Name of feature	Remark/key point
Time Synchronous Averaging based features extraction		
1	TSA	$y_{TSA} = \frac{1}{N}\sum_{i=0}^{N-1} y(t+nT)$, where, T= period of averaging, N = number of sample points
2	NA4	$NA4 = \frac{N\sum_{i=1}^{N}(y_i-\bar{y})^4}{\left\{\frac{1}{M}\sum_{j=1}^{M}\left[\sum_{i=1}^{N}(y_{ij}-y_j^2)\right]\right\}^2}$, where, M = number of record in current time, N = total number of sample
3	NA4*	$NA4^* = \frac{N\sum_{i=1}^{N}(y_i-\bar{y})^4}{\left(\sigma_{y-h}^2\right)^2}$, where, $\left(\sigma_{y-h}^2\right)$ = variance
4	FM4	$FM4 = \frac{N\sum_{i=1}^{N}\left(d_i-\bar{d}\right)^4}{\left[\sum_{i=1}^{N}\left(d_i-\bar{d}\right)^2\right]^2}$, where, $\bar{d}$ = mean of d = difference signal
5	M6A	$M6A = \frac{N^2\sum_{i=1}^{N}\left(d_i-\bar{d}\right)^6}{\left[\sum_{i=1}^{N}\left(d_i-\bar{d}\right)^2\right]^3}$
6	M8A	$M8A = \frac{N^3\sum_{i=1}^{N}\left(d_i-\bar{d}\right)^8}{\left[\sum_{i=1}^{N}\left(d_i-\bar{d}\right)^2\right]^4}$
Time series regressive based features extraction methods		
1	AR (p)	$y_t = a_1y_{t-1}+a_2y_{t-2}+....+a_py_{t-p}+\mu_t$, and $y_t = \mu_t+\sum_{i=1}^{p}a_iy_{t-i}$, where, μ_t = noise ratio, a1-ap = model parameters, p = model order
2	MA (q)	$y_t = b_1\mu_{t-1}+b_2\mu_{t-2}+....+b_q\mu_{t-p}+\mu_t$ and $y_t = \mu_t+\sum_{i=1}^{q}b_i\mu_{t-i}$ where, b1-bp = model parameters, q = model order

TABLE 1.1 Brief summary of TDbFE for CM of smart grid applications. (*Cont.*)

S. no.	Name of feature	Remark/key point
3	ARMA (p.q)	$y_t = a_1 y_{t-1} + \ldots + a_p y_{t-p} + \mu_t + b_1\mu_{t-1} + \ldots + b_q\mu_{t-p}$ and $y_t = \mu_t + \sum_{i=1}^{p} a_i y_{t-i} + \sum_{i=1}^{q} b_i \mu_{t-i}$
4	ARIMA (p,D,q)	$\Delta^D y_t = a_1\Delta^D y_{t-1} + \ldots + a_p\Delta^D y_{t-p} + \mu_t + b_1\mu_{t-1} + \ldots + b_q\mu_{t-p}$, where Δ^D = difference
Statistical features extraction		
1	Minimum value	$\min[y_i]$, where y = input signal data from the sensor
2	Mean value	$\bar{y} = \frac{1}{N}\sum_{i=1}^{N} y_i$, where, $\bar{y}$ = mean value
3	Maximum	$\max[y_i]$
4	Root-mean-squire value	$y_{RMS} = \sqrt{\frac{1}{N}\sum_{i=1}^{N} y_i}$
5	Peak-to-peak value	$y_{PtP} = [y_i(\max) - y_i(\min)]$
6	Crest factor	$CF = \frac{y_{PtP}}{y_{RMS}}$
7	Variance	$\mathrm{var}_y = \sigma_y^2 = \frac{\sum (y_i - \bar{y})^2}{N-1}$
8	Standard deviation	$STD = \sigma_y = \sqrt{\frac{\sum (y_i - \bar{y})^2}{N-1}}$
9	Standard error value	$y_{STE} = \sqrt{\frac{1}{N-1}\left[\sum (y_i - \bar{y})^2 - \frac{\left[\sum (x-\bar{x})(y-\bar{y})\right]}{\sum (x-\bar{x})^2}\right]}$, where, x = predicted value
10	Zero crossing	Number of *x*-axis crossing with satisfy the two criteria: $y_i > 0$ *and* $y_{i-1} < 0$ and $y_i < 0$ *and* $y_{i-1} > 0$
11	Wavelength	$y_{WL} = \sum_{i=1}^{N} \lvert y_i - y_{i-1} \rvert$
12	Willison amplitude value	$y_{WA} = \sum_{i=1}^{N} f(\lvert y_i - y_{i+1} \rvert)$, if s.t $f(y) = \begin{cases} 1 & \text{if } y \geq \tau \\ 0 & \text{otherwise} \end{cases}$ and τ = threshold

(*Continued*)

TABLE 1.1 Brief summary of TDbFE for CM of smart grid applications. (*Cont.*)

S. no.	Name of feature	Remark/key point
13	Impulse factor	$y_{IF} = \dfrac{y_{peak}}{\frac{1}{N}\sum_{i=1}^{N} \lvert y_i \rvert}$
14	Margin factor	$y_{MF} = \dfrac{y_{peak}}{\left(\frac{1}{N}\sum_{i=1}^{N} \sqrt{\lvert y_i \rvert}\right)^2}$
15	Shape factor	$y_{SF} = \dfrac{y_{RMS}}{\frac{1}{N}\sum_{i=1}^{N} \lvert y_i \rvert}$
16	Clearance factor	$y_{CLF} = \dfrac{y_{\max}}{\left(\frac{1}{N}\sum_{i=1}^{N} \sqrt{\lvert y_i \rvert}\right)^2}$
17	Skewness	$y_{SK} = \dfrac{\sum_{i=1}^{N} (y_i - \bar{y})^3}{N\sigma_y^3}$
18	Kurtosis	$y_{KR} = \dfrac{\sum_{i=1}^{N} (y_i - \bar{y})^4}{N\sigma_y^4}$
19	Histogram	$H_{LB} = y_{\min} - 0.5\left(\dfrac{y_{\max} - y_{\min}}{N-1}\right)$, where LB = lower bound $H_{UB} = y_{\max} - 0.5\left(\dfrac{y_{\max} - y_{\min}}{N-1}\right)$, where UB = upper bound
20	Normal PDF	$N_{PDF} = \dfrac{1}{\sigma\sqrt{2\pi}} \exp^{-\left(y_i - \frac{\mu}{2\sigma^2}\right)^2}$ where, PDF
21	Weibull PDF	$W_{PDF} = \dfrac{b}{a}\left(\dfrac{y_i}{a}\right)^{b-1} \exp^{-\left(\frac{y_i}{a}\right)}$
22	Negative Log-Likelihood	$-\log L = -\sum_{i=1}^{N} \log\left[f(a, b \setminus y_i)\right]$
23	Entropy	$y_{ENT} = \sum_{i=1}^{N} P_{y_i} \log P_{y_i}$, where, P_{y_i} = probabilities computed from the distribution of y.

TABLE 1.1 Brief summary of TDbFE for CM of smart grid applications. (*Cont.*)

S. no.	Name of feature	Remark/key point
Filter based features extraction methods		
1	Demodulation	Demodulation of a signal (DoS) process includes three basic steps: (1) pass the signal through band-pass filter, (2) apply HHT (Hilber-Huang transform), and (3) apply FFT (fast Fourier transform)
2	Prony method	Prony model is similar to AR, ARMA, ARIMA, and try to fit the model with respect to the signal. Prony, $\hat{y}(t)=\sum_{i=1}^{L} A_i e^{-\sigma_i t}\cos(w_i t+\phi)$, where L = damped exponential component, A_i = amplitude, w = angular frequency, σ_i = damping coefficient and ϕ = phase shift
3	Adaptive noise cancellation (ANC)	ANC remove the background noise of the waveform. For this, reference signal is passed though the adaptive filter then subtract from original signal y(t) to obtain the error signal. Then error signal is feedback to adaptive filter (with adjustment of filter parameters) to minimize the error output

TABLE 1.2 Brief summary of FDbFE for CM of smart grid applications.

S. no.	Name of feature	Remark/key point
Fourier analysis based features extraction methods		
1	Fourier series analysis (FSA)	The FSA is a harmonic analysis of a signal. It decomposes the signal into sinusoidal components, which includes amplitude and phase. Based on these amplitude and phase, further analysis for CM is performed
2	Discrete Fourier transform (DFT)	The DFT is performed to decompose the signal for further analysis. For detail, MATLAB user guide may be referred (Specially Communication System Toolbox)
3	Fast Fourier transform (FFT)	FFT is an efficient way to evaluate the DFT. For detail, MATLAB user guide may be referred (Fourier analysis and filtering package)
Frequency spectrum statistical features based features extraction methods		
1	Arithmetic mean	$\hat{y}_{AM}(w)=20\log\left[\frac{\frac{1}{N}\sum_N A_n}{10^{-5}}\right]$, where, A_n = amplitude

(*Continued*)

TABLE 1.2 Brief summary of FDbFE for CM of smart grid applications. (*Cont.*)

S. no.	Name of feature	Remark/key point
2	Geometric mean	$\hat{y}_{GM}(w) = \sum_N 20\log\left[\frac{\frac{A_n}{\sqrt{2}}}{10^{-5}}\right]$, it is in dBm representation
3	Matched filter RMS	$M_{fRMS} = 10\log\left[\frac{1}{N}\sum_N\left(\frac{A_i}{A_n^{ref}}\right)\right]$
4	RMS of spectral difference	$R_{dRMS} = \sqrt{\frac{1}{N}\sum_N\left(P_n - P_n^{ref}\right)^2}$
5	Sum of squares spectral difference	$R_{dSD} = \frac{1}{N}\sum_N\sqrt{\left(P_n - P_n^{ref}\right) * \left\lvert P_n - P_n^{ref}\right\rvert}$
Envelope analysis (EA)	The EA is known as HFRA (high-frequency resonance analysis or resonance modulation). Mostly, it is used in rotating machine fault identification. The process of EA is similar to the DoS process and includes three basic steps: (1) pass the signal through band-pass filter, (2) apply HHT (Hilber-Huang transform), and (3) apply FFT (fast Fourier transform)	
Other features extraction		
1	Mean frequency	$f_{MF} = \frac{\sum_{i=1}^{N} f_i^2 \cdot S(f_i)}{\sum_{i=1}^{N} S(f_i)}$, where, N = number of spectrums
2	Average frequency	$f_{AvgF} = \frac{\sum_{i=1}^{N} f_i^4 \cdot S(f_i)}{\sum_{i=1}^{N} f_i^2 \cdot S(f_i)}$, It is the wave shape of signal crosses the mean of time-domain signal

TABLE 1.2 Brief summary of FDbFE for CM of smart grid applications. (*Cont.*)

S. no.	Name of feature	Remark/key point
3	Stabilization factor of wave shape	$WaveShape_{SF} = \frac{\sum_{i=1}^{N} f_i^2 \cdot S(f_i)}{\sqrt{\sum_{i=1}^{N} S(f_i) \sum_{i=1}^{N} f_i^4 \cdot S(f_i)}}$
4	Coefficient of variability	$Cof_v = \frac{\sigma}{\bar{f}}$, where, $\sigma = \sqrt{\frac{\sum_{i=1}^{N} \sqrt{\left(f_i - \bar{f}\right)^2} \cdot S(f_i)}{N}}$
5	Frequency-domain skewness	$f_{SK} = \frac{\sum_{i=1}^{N} \left(f_i - \bar{f}\right)^3 \cdot S(f_i)}{\sigma^3 N}$, where $\bar{f} = \frac{\sum_{i=1}^{N} (f_i) \cdot S(f_i)}{\sum_{i=1}^{N} S(f_i)}$
6	Frequency-domain kurtosis	$f_{KR} = \frac{\sum_{i=1}^{N} \left(f_i - \bar{f}\right)^4 \cdot S(f_i)}{\sigma^4 N}$
7	Root-mean-square ratio	$f_{RMSR} = \frac{\sum_{i=1}^{N} \sqrt{\left(f_i - \bar{f}\right)} \cdot S(f_i)}{\sqrt{\sigma} \cdot N}$
	Wavelength	$y_{WL} = \frac{WaveSpeed}{f}$
	Zero-crossing	$f = \frac{F_s}{NumberCycles}$, where, F_s = sampling frequency

TABLE 1.3 Brief summary of TFDbFE for CM of smart grid applications.

S. no.	Name of feature	Remark/key point
Fourier transform based analysis		
1	Short-time Fourier transform (STFT)	STFT is the first modified version of FT (Fourier transform), which is utilized for decomposition of non-stationary signals in time-frequency domain. For complete mathematical implementation, signal processing toolbox may be referred.

(*Continued*)

TABLE 1.3 Brief summary of TFDbFE for CM of smart grid applications. (*Cont.*)

S. no.	Name of feature	Remark/key point
Wavelet analysis based analysis		
1	Continuous wavelet transform (CWT)	CWT is a wavelet transform (WT), which is represented as: $CWT_{y(t)}(s,\tau)=\frac{1}{\sqrt{s}}\int y(t)\psi^*\left(\frac{t-\tau}{s}\right)dt$, where, ψ^* = complex conjugate of $\psi(t)$ and (s,τ) = parameters. For complete mathematical implementation, Wavelet toolbox may be referred.
2	Discrete wavelet transform (DWT)	DWT is discrete form of CWT. Here, $\psi_{s,\tau}(t)$ is discretized by utilizing dynamic scales ($s=2^j$ and $\tau=k2^j$), here j and k are integers. Normally, DWT is developed by using low-pass and high-pass filters. For complete mathematical implementation, wavelet toolbox may be referred.
3	Wavelet packet transform (WPT)	WPT is the improvement of DWT. For complete mathematical implementation, wavelet toolbox may be referred.
Empirical mode decomposition (EMD) based analysis		EMD decompose the signal into time-frequency domain. It generates the IMF like frequency band (FB) in wavelet. The procedure of EMD is as follow: 1. Find out the local extrema and then connect by using cubic spline as upper envelop. 2. Repeat this process for local minima to create lower envelop. And both envelop should cover all data samples. 3. Evaluate mean value of both envelop and then find difference (h1) from main signal y(t). If h1 is IMF then h1 is first component of y(t), else h1 treated as original signal and same procedure will be followed till residue value of the signal. For more information of EMD, Chapter 5 may be referred.
Hilbert-Huang transform (HHT)		The HHT can compute the instantaneous frequency of a signal, which is represented by: $HHT_{y(t)}=\frac{p}{\pi}\int_{-\infty}^{+\infty}\frac{y(\tau)}{t-\tau}d\tau$ where, p = principal value of the singular integral. For complete mathematical implementation, MATLAB user manual may be referred
Wigner-Ville distribution (WD)		The WD for a signal y(t) can be represented by: $W_y(t,f)=\int_{-\infty}^{+\infty}y\left(t+\frac{\tau}{2}\right)y^*\left(t+\frac{\tau}{2}\right)e^{-2\pi ft}d\tau$, where y^* = complex conjugate of $y(t)$. For complete mathematical implementation, MATLAB user manual may be referred
Local mean decomposition (LMD)		LWD is an adaptive technique for data decomposition in form of product functions (PFs). For complete mathematical implementation, MATLAB user manual may be referred

The list of different versions of the wavelet (i.e., 65 versions of 11 categories) can be chosen for the TFbFE in the study are presented as:

1. Haar wavelet (haar: 1 case).
2. Daubechies wavelets (db1 to db10: 10 different cases).
3. Symlets wavelets (sym2 to sym8: 7 cases).
4. Coiflets wavelets (coif1 to coif5: 5 cases).
5. Bi-orthogonal wavelets (bior1.1 to bior6.8: 15 cases).
6. Reversed Bi-orthogonal wavelets (rbio1.1 to rbio6.8: 15 cases).
7. Meyer wavelet (meyr: 1 case).
8. Discrete approximation of Meyer wavelet (dmey: 1 case).
9. Gaussian wavelets (gaus1 to gaus8: 8 cases).
10. Mexican hat wavelet (mexh: 1 case).
11. Morlet wavelet (morl: 1 case).

In summary, FEP has been implemented in each chapter of the book separately. So, with consolation of not to repeat same information twice, the feature extraction approach has been explained and explained in detailed in the subsequence chapters.

3.2 Most relevant features selection

Feature selection is an approach of selecting a most relevant subset features for developing a robust AI/machine learning model. In the feature selection process, redundant and/or irrelevant data is removed from the main database, hence performance of the diagnostics model may be improved. By feature selection process, computational burden on the machine will be reduced, hence computational efficiency is increased.

Generally, in future selection procedure four main steps should be involved as given: (1) subset generation; (2) evaluation of the subset; (3) procedure stopping criteria; and (4) validation. In step#1, sub-sets are selected based on search approach. Generally, approach depends on (1) search direction and (2) search methodology. Step#2 depends on several evaluation parameters such as distance, dependency, consistency, etc. In the step#3, stopping criteria depends on several other criteria (i.e., error is less than required/chosen, complete the search, etc.), and in step#4, the validation of selected attribute is performed using different advanced AI/ML algorithms.

There are different types of feature selection methods. Some of them are listed in Table 1.4. There are three main categories of feature selection approaches. These approaches are: (1) filter-based feature selection (FbFS); (2) wrapper-based feature selection (WbFS); and (3) embedded model based feature selection (EMbFS). Generally, FbFS is the preprocessing step of any AI/machine learning approach. FbFS method measures the quality (correlation, similarity, information, and dependency) of dataset. Due to this, FbFS methods are fast and free from high computational calculations. The WbFS

TABLE 1.4 Brief summary of feature selection methods for smart grid applications.

S. no.	Name of feature	Remark/key point
Filter based feature selection		
1	Fisher score (FS)	FS is to select the feature of supervised type of data.
2	Laplacian score (LS)	LS is the unsupervised filter based approach.
3	Relief and Relief-F algorithms (R&RF)	R&RF methods are supervised type of techniques Relief algorithm is a binary class algorithm Relief-F algorithm is the updated version of Relief algorithm. It can handle data with noise signal, incomplete and multiclass.
4	Pearson correlation coefficient (PCC)	PCC is a supervised method of feature selection. It work based on ranking search. PCC evaluate the relation between two variables as: $r = \frac{\text{cov}(p_i, q)}{\sqrt{\text{var}(p_i) * \text{var}(q)}}$, where $-1 \leq r \geq 1$ negative sign or r shows the inverse correlation and so on.
5	Information gain (IG) and gain ratio (GR)	It is work based on concept of entropy. It also overcomes the limitation of PCC as a linear correlation measure.
6	Mutual information (MI)	MI evaluate dependency in between two variables, which is evaluated as: $MI\big\|_{(p.q)} = \sum_{p \in P} \sum_{q \in PQ} \rho(p,q) \log \frac{\rho(p,q)}{\rho(p)\rho(q)}$
7	Chi-squared (Chi-2)	The ranking procedure is based on the χ^2 test statistic, which is computed as: $\chi^2(f,c) = \frac{L\left(E_{c,f}E - E_c E_f\right)^2}{\left(E_{c,f} + E_c\right)\left(E_f + E\right)\left(E_{c,f} + E_f\right)\left(E_c + E\right)}$, where, L = number of example, $E_{c,f}$ = number of repetition of c and f, E_f = repeated occurrence of f without c, E_c = repeated occurrence of c without f, E = repeated occurrence of f nor c. big value of χ^2 represents the highly correlation.
8	Wilcoxon ranking (WR)	WR is a nonparametric test used to evaluate ranking of features.
Wrapper model–based feature subset selection		
1	SSA	There are two main type of SSA: (1) sequential forward selection (SFS) and (2) sequential backward selection (SBS). These methods are SFFS (sequential forward floating selection) and SBFS (sequential backward floating selection)

TABLE 1.4 Brief summary of feature selection methods for smart grid applications. (*Cont.*)

S. no.	Name of feature	Remark/key point
2	HbSA	The HbSA includes several advanced AI/ ML approaches such as: (1) ACO (ant colony optimization), PSO (particle swarm optimisation), GA (genetic algorithms), and GP (genetic programming), etc.
Embedded model–based feature selection (EMbFS)		
1	CART	All classification and regression tree
2	C4.5	Like, J48 algorithm
3	SVM	SVM-RFE, MSVM, etc.

is based on predictive performance of predefined predictor. WbFS is not economic in comparison to FbFS methods. Some of the approaches under WbFS methods are: (1) Sequential Selection Algorithms (SSA) and (2) Heuristic-Based Selection Algorithms (HbSA). In the EMbFS category of feature selection, the following approaches are come in the list as: LASSO, classification & regression tree (CART), C4.5, and SVM-RFE, MSVMs, etc. In the summary, feature selection procedure has been implemented in each chapter of the book separately. So, in order to remove the repetition of the text/information, feature selection approach has been explained in the sub-sequence chapters.

4 Data visualization and correlation representation (DVCR)

Data visualization is a graphical representation of the data, so that user can understand in an easy way. It includes the generating/creating the plots that represents the relationship among presented dataset to the user of the plots. Such type of communication is achieved through the systematic mapping of the data value in the plots. To communicate the important information in clear way, data visualization uses the statistical tool, graphics, etc., and some of them are listed as: (1) line plots (i.e., error bar, plot, semilogx, semilogy, loglog, area, comet, stacked plot); (2) stem, stem3, and stair plots; (3) bar plots (bar, barh, bar3, bar3h, histogram and pareto, barstacked, plot matrix); (4) scatter plots (i.e., spy, simple scatter, scatter3, plotmatrix, etc.); (5) pie chart (i.e., 2-d and 3-d); (6) polar plots (i.e., polar, rose, compass); (7) vector fields (feather or compass type); (8) analytics plots; (9) signal processing (window based visualization); (10) signal processing: spectral estimation based (i.e., psd, power, psd-pwelch, power-pwelch, pmtm, pburg, pcov, pmcov, pyulear, pelg, pmusic, spectrogram, strips, etc.); (11) machine learning based plots (i.e., box plot, ecdf plot, histfit plot, ksdensity, normplot, Weibull plot, QQ plot, Andrews plot, glyphplot, etc.); (12) graph plot; (13) geographic plots (i.e., geoshow, geoplot, geoscatter, geobubble, etc.); (14) contour plots (i.e., simple contour, contour, contour3 plot); (15) image plots (i.e., simple image, pcolor, imagesc, imshow, heatmap); (16) three-D surfaces (i.e., surf, surfc, surfl, mesh, meshz, waterfall, ribbon, contour3 plots); (17) financial plots (i.e., highlow, candle, kagi, renko, priceandvol, volarea, pointfig plot); and (18) filter base plots (i.e., fvtool, magnitude, phase, freqs, impz, stepz, zphane, etc.).

The effective visualization helps the user to analyze the plot according to the data points. Hence, DVCR is an art as well a science. Increasing amount of data coming from IoTs nodes and/or sensors nodes create a big data for the analysis. So, due to high volume of data, there is a challenge for visualizing such big data in a systematic way. Therefore, DVCR analysis is another research area in the intelligent data analytics, which can address the several important problems. There are numerous tools and techniques for DVCR as mentioned in the Section 6.

There are several visualization technologies as listed: Ajax, GIS, Model-View-Controller (MVC) architecture, Software Development Kits (SDKs), Java 2D, Swing APIs, Styling and Data Mapping (SDM), DMS latitude/longitude, decimal degrees, UTM or MGRS, Vector, Raster, Geocentric, Geodesic, Decimal Lat/Lon.

To govern the technologies, there are some visualization standards as listed: NATO APP-6A/MIL-STD 2525B, ESRI, GeoTIFF, "DTED 0, 1 & 2," GTOPO30 DEM, MapInfo, Oracle Spatial, TIGER/Line, DMS Lat/Lon, UTM, GRS 80, North American 1983 (CONUS), Geodetic, NIMA/NGA, XML, Digest, Java2, XML, DIGEST, NIMA, RPF, VPF, and SBGN, etc.

5 Application area

The application area of the AI/machine learning (ML) and intelligent data analytics is very vast. In the modern field, everywhere is AI/ML because ML algorithms can learn automatically through the experience. According to the focused area of CM of smart grid application, following main area of implementation have been focused in this study. These areas are: (1) Clustering; (2) Regression; (3) Classification; and (4) Forecasting. As per the application of smart grid, the classification of AI/ML algorithm and its associated implementation have been represented in Fig. 1.2. Fig. 1.2 comprises of the all type of ML algorithm and its uses along with labeled and unlabeled dataset. Thereafter, Fig. 1.2 is represented as the zoom level of the Fig. 1.1 from algorithm point of view. At least one case study based on these listed application area has been presented in this book. For example, AI/ML and data analytic application in anomaly detection have been presented in Chapter 2 (for PV fault diagnosis), Chapter 3 (for power transformer fault diagnosis), Chapter 4 (for IM fault diagnosis), Chapter 5 (for power quality disturbance diagnosis), Chapter 6 (for transmission line fault

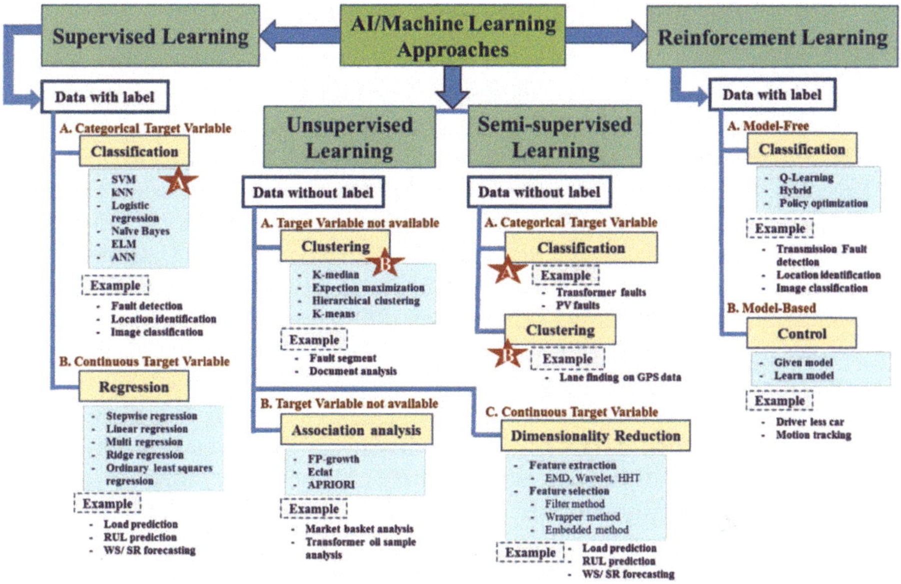

FIGURE 1.2 **Representation off AI/machine learning approaches for different application area.**

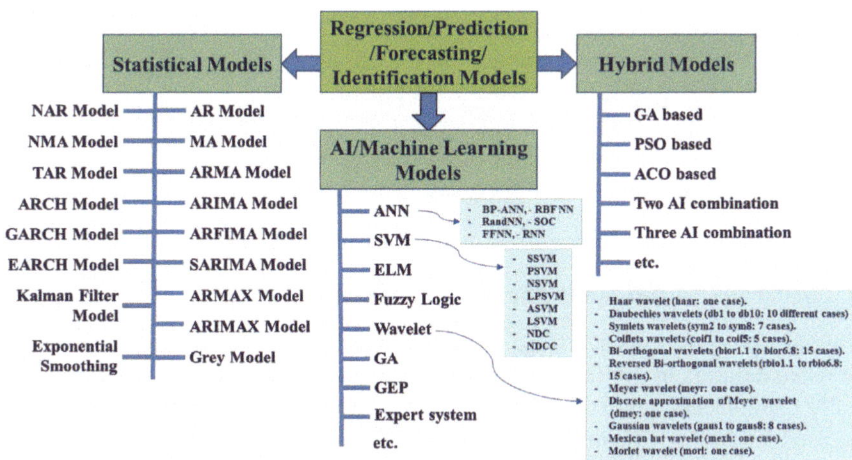

FIGURE 1.3 **Conventional and advance AI/machine learning approaches for different application area.**

diagnosis). Similarly, AI/ML applications for prediction, forecasting have been presented in Chapter 7 (for SR forecasting), Chapter 8 (for WS forecasting), Chapter 9 (for load forecasting), and Chapter 10 (for battery's remaining useful life forecasting and monitoring) (Fig. 1.3).

6 Softwares and techniques used for data analytics

There are numerous data visualization softwares, which can be utilized for the further analysis point of view. Out of these software/tool, few are open access. All these tools and software are listed as follows: (1) Amira (software), (2) AnyChart, (3) Aphelion (software), (4) Avizo (software), (5) CartoDB, Chart.js, (6) DADiSP, (7) Datacopia, (8) DataScene, (9) DataViva, (10) Datawatch Corporation, (11) Domo (company), (12) EMovie, (13) Endrov, (14) Eye-Sys, (15) FICO, (16) FlexPro, (17) FusionCharts, (18) Gliffy, (19) GoodData, (20) Google Public Data Explorer, (21) GRAPE, Highcharts, (22) Histcite, (23) HotSauce, (24) IBM OpenDX, (25) ILNumerics, (26) Infogram, (27) InfoZoom, (28) Jedox, (29) JGraph, (30) JMP (statistical software), (31) Kibana, (32) Kitware, (33) LIONsolver, (34) BioTapestry, (35) Cytoscape, (36) GenMAPP, (37) MEGA, (38) PathVisio, (39) InCroMAP, (40) Pathview, (41) Systrip, (42) GESTALT Workbench, (43) N-Browse, (44) NetPath, (45) REACTOME, (46) WikiPathways, (47) MetaboMAPS, (48) Looker (company), (49) Maple (software), (50) Mathcad, (51) MATLAB, (52) MeVisLab, (53) Microsoft Power BI, (54) MicroStrategy, (55) Molecular, (56) Evolutionary Genetics Analysis, (57) Moodbar, (58) Orange (software), (59) OSIsoft, (60) Our World in Data, (61) PathVisio, (62) Pipeline Pilot, (63) Plotly, (64) Pyramid Analytics, (65) Qlik, (66) Qunb, (67) RGraph, (68) RJMetrics, (69) RW3 Technologies, (70) Seeq Corporation, (71) Spotfire, (72) Starlight Information Visualization System, (73) Studierfenster, (74) T-REX (webserver), (75) Tableau Software, (76) Targit (company), (77) Teechart, (78) Tom Sawyer Software, (79) Tomviz, (80) Trade Space Visualizer, (81) Trendalyzer, (82) Vaa3D, (83) Visual.ly, (84) Volume cartography, (85) Wolfram Language, etc.

After the data visualization and correlation representation, data analytics is performed using different ML algorithms based on different software and tools. Some software and tools out of them are listed here (Table 1.5) for the easiness of the reader.

7 Sources of datasets for data analytics

This section comprises a vast list of available datasets in the digital domain. Out of listed datasets in the Table 1.6, some datasets are open access, some are partially open access, and some datasets have a copy right. So, these dataset may be used for further research and industry application as per the copy right of the dataset. Apart of these dataset, several other datasets have also been presented in the different chapters of this book. For example, a big list of dataset and available tools and software related to renewable energy resource assessment (RES data analytics) has been presented in Chapter 8.

8 Conclusion

In this chapter, introduction related to AI, ML, and intelligent data analytics for the CM of smart grid application has been presented. In the introduction part, following key points have been represented: (1) brief information AI,

TABLE 1.5 Summary of machine learning based data analytics software and tools.

S. no.	Name of machine learning software or tool	URL		Written platform
Open source software				
1	Caffe	caffe.berkeleyvision.org		C++
2	CNTK	https://docs.microsoft.com/en-us/cognitive-toolkit/	Microsoft Research	C++
3	Deeplearning4j	https://deeplearning4j.org/	Various	Java, CUDA, C, C++
4	DeepSpeed	https://www.deepspeed.ai/	Microsoft	Python, CUDA, C++
5	ELKI	elki-project.github.io	Technical University of Dortmund; initially Ludwig Maximilian University of Munich	Java
6	Infer.NET	https://dotnet.github.io/infer/	Microsoft, .NET Foundation	C#
7	Keras	https://keras.io/	Various	Python
8	Mahout	https://mahout.apache.org/	Apache Software Foundation	Java, Scala
9	Mallet	http://mallet.cs.umass.edu/	Andrew McCallum et al.	Java
10	ML.NET	https://dotnet.microsoft.com/apps/machinelearning-ai/ml-dotnet	.NET Foundation	C# and C++
11	XGBoost	https://xgboost.ai/	The XGBoost Contributors	C++, Python, Java, R
12	LightGBM	https://lightgbm.readthedocs.io/en/latest/	Microsoft and LightGBM Contributors	C++, Python, R, C
13	mlpack	https://mlpack.org/		C++, Python, Julia, Go

(Continued)

TABLE 1.5 Summary of machine learning based data analytics software and tools. (*Cont.*)

S. no.	Name of machine learning software or tool	URL		Written platform
14	MXNet	mxnet.apache.org	Apache Software Foundation	C++, Python, R, Java, Julia, JavaScript, Scala, Go, Perl
15	Neural Lab	http://www.fimee.ugto.mx/profesores/sledesma/documentos/		C
16	Octave	https://www.gnu.org/software/octave/	John W. Eaton et al.	C, C++, Fortran
17	OpenNN	https://www.opennn.net/	Artelnics	C++
18	Orange	https://orange.biolab.si/	University of Ljubljana	Python, Cython, C++, C
19	ROOT (TMVA with ROOT)	https://root.cern/	CERN	C++
20	scikit-learn	https://scikit-learn.org/stable/	David Cournapeau	Python, Cython, C and C++
21	Shogun	https://www.shogun.ml/	Soeren Sonnenburg et al.	C++
22	Spark MLlib	https://spark.apache.org/	Apache Spark	Scala
23	SystemML	http://systemds.apache.org/	Apache Software Foundation, IBM	Java
24	TensorFlow	https://www.tensorflow.org/	Google Brain Team	Python, C++, CUDA
25	Torch/PyTorch	http://torch.ch/ https://pytorch.org/	Ronan Collobert et al. Facebook's AI Research lab (FAIR)	Lua, LuaJIT, C, CUDA and C++
26	Weka/MOA	https://www.cs.waikato.ac.nz/~ml/weka/	University of Waikato	Java

27	Yooreeka	https://code.google.com/archive/p/yooreeka/	Haralambos Marmanis	Java
28	pandas (software)	https://pandas.pydata.org/	Community	Python, Cython, C
Proprietary software with open source editions				
1	KNIME	https://www.knime.com/	KNIME	
2	RapidMiner	https://rapidminer.com/	RapidMiner	
Proprietary software				
1	Amazon Machine Learning	https://aws.amazon.com/		
2	Angoss Knowledge STUDIO	https://www.altair.com/knowledge-studio/		
3	Azure Machine Learning	https://azure.microsoft.com/en-us/	Microsoft	
4	Ayasdi	https://www.ayasdi.com/	Ayasdi	
5	IBM Data Science Experience	https://www.ibm.com/sg-en	IBM Corp.	R/Python/Scala
6	Google Prediction API	www.google.com	Google	Java, Python
7	IBM SPSS Modeler	https://www.ibm.com/products/spss-modeler	IBM Corp.	
8	KXEN Modeler	https://www.sap.com/index.html	KXEN Inc.	
9	LIONsolver	https://intelligent-optimization.org/	Reactive Search srl	
10	Mathematica	https://www.wolfram.com/mathematica/	Wolfram Research	Wolfram Language, C/C++, Java
11	MATLAB	https://www.mathworks.com/products/matlab.html	MathWorks	C/C++, MATLAB

(Continued)

TABLE 1.5 Summary of machine learning based data analytics software and tools. (*Cont.*)

S. no.	Name of machine learning software or tool	URL		Written platform
12	Neural Designer	https://www.neuraldesigner.com/	Artelnics	C++
13	NeuroSolutions	http://www.neurosolutions.com/	NeuroDimension	
14	Oracle Data Mining	https://www.oracle.com/database/technologies/advanced-analytics/odm.html	Oracle Corporation	
15	Oracle AI Platform Cloud Service			
16	RCASE		Warwick University	
17	SAS Enterprise Miner	https://www.sas.com/en_us/home.html	SAS Institute	C
18	SequenceL	http://www.texasmulticore.com/	Texas Tech University, Texas Multicore Technologies	
19	Splunk	https://www.splunk.com/	Public	
20	STATISTICA Data Miner	https://www.tibco.com/products/data-science	TIBCO Software	

TABLE 1.6 Summary of different level dataset to be used for data analysis.

S. no.	Name of provider	Remarks
(a) Country Databases		
1	EIU.com	- It provides holistic coverage - It provides ~41 country reports - It provides ~16 country forecasts
2	Passport	- It includes consumer reports for most countries - It includes consumer trends database
3	World Bank	- It has statistical country profiles - It presents various global indicators - It is the open access database
(b) Industry level databases		
1	Passport	- It provides the consumer products information - Some retail services information - It provides broad range of statistical databases
2	Frost & Sullivan	- It has broad range of industries databases - It focus on tech advantages - Most of report are in similar patters
3	Fitch Connect	- It covers specific range of industry and its data - It shorts highlights daily basis - It also provide quarterly reports
4	Gartner	- It covers IT sector in various industry - It provides hype cycles as well
(c) Company data bases		
1	Orbis	- It provides listed and private company globally - It provides in-depth financial information
2	D&B Hoovers	- It also provides full list of listed and private company globally - It shows the corporate family structure
3	Factiva	- It has mostly listed companies of the globe - It provides the company snapshots, Headlines and Peer Lists
4	Patsnap	- It provides invention data from the globe related to all research domain
5	Lens.org	- It also provides invention data - Open access database
(d) Some other resources		
1	Economist Intelligence Unit (EIU)	Blogs and free reports from major market research companies
2	Euromonitor	
3	McKinsey	
4	Mintel	
5	Enterprise Singapore	- Offers free statistics and analysis
6	Export.gov	- The United States website but with international data
7	Industry.gov.au	- Offers free statistics and analysis

(*Continued*)

TABLE 1.6 Summary of different level dataset to be used for data analysis. (*Cont.*)

S. no.	Name of provider	Remarks
8	National Library	- NLB has some free or partially free databases
9	UNIDO Statistics Data Portal	- Some free or partially free databases
10	Statista	- Some free or partially free databases - Can use free version
11	IMF e-Library	- Some free or partially free databases
(e) Open access data bases		
(1) AGRIS, (2) Art in context, (3) Air University Library index to military periodicals, (4) Art in context, (5) arXiv.org, (6) Asian Shakespeare intercultural archive, (7) Bentham Open BioMed Central the open access publisher (Open Access), (8) The Biodiversity Library of Southeast Asia (BLSEA) (BLSEA), (9) The birds of Singapore, (10) ChemSpider, (11) Cinemas of Asia journal of the Network for the Promotion of Asian Cinema, (12) CML CMI database, (13) Colonial film moving images of the British empire, (14) CumInCAD cumulative index of computer aided architectural design, (15) Contemporary Wayang Archive, (16) China Judgements Online, (17) Digital Commons network, (18) E-STAT, (19) E.stat for Business and Economics, (20) eAtlas of Global Development, (21) Economics Research Network, (22) Electronic Theses & Dissertations (ETD), (23) Europe Pubmed Central, (24) FAO catalogue online, (25) Financial Economics Network, (26) Government of Canada: Depository Services Program (DSP) E-Collection, (27) Grey Literature Network Service, (28) The Global Competitiveness Report, (29) Contemporary Chinese Area Studies, (30) IMF data, (31) INIS International Nuclear Information System, (32) Journal of Entrepreneurship, (33) Management and Innovation, (34) Kinokuniya BookWEB Pro, (35) The Kyushu University Museum Digital Archive, (36) Center for Southeast Asian Studies, (37) Kyoto University (CSEAS), (38) Japan Center for Asian Historical Records, (39) The National Diet Library Digital Archive Portal, (40) Southeast Asia, LearningIn10.com, (41) Lee Kong Chian Natural History Museum Ebooks, (42) Legal Scholarship Network, (43) Management Research Network, (44) MCI press room, (45) MEDLINEplus, (46) MICA press releases and speeches, (47) Monthly Digest of Statistics, (48) Neliti Indonesia's Research Repository, (49) NewspaperSG, (50) Resource Sharing System for the Humanities, (51) OAPEN, (52) Official Document System of the United Nations, (53) OKR Open Knowledge Repository, (54) Open Textbook Library, (55) PubChem, (56) PubMed, (57) QALAM Article Database, (58) Research Catalogue-an international database for artistic research (Society for Artistic Research, Open Access), (59) Research IRM, (60) Rhetoric & Communication Research Network, (61) RIBA Library Online Catalogue, (62) Singapore Government Directory, (63) Singapore Research Nexus, (64) SingStat Table Builder (Open Access), (65) Social Science Research Network SSRN, (66) Southeast Asia in the Ming Shi-lu: An Open Access Source, (67) The scientific crucible science education in NUS since 1929, (68) Kobe University Library Digital Archive : Newspaper Article, (69) The Japan Society for Southeast Asian Studies (JSSEAS), (70) Network for Southeast Asian Studies, (71) The Japanese Bibliography of Southeast Asian Studies (JABSEAS), (72) National Digital Library of Theses and Dissertations in Taiwan, (73) United Nations Treaty Collection, (74) Wayang Kontemporer Innovations in Javanese Wayang Kulit, (75) Wilson Center Digital Archive International History Declassified, (76) World Bank Open Data, (77) World Development Indicators, (78) World Economic Forum, (79) WSH Council, (80) YNUJ: the Yale-NUS Undergraduate Journal, (81) MyManuskrip Digital library of Malay Manuscript (Pustaka Digital Manuskrip Melayu).		

ML, and data analysis, (2) data analytics for business, (3) data analytics for smart grid, and (4) data analytics for CM. Brief detail of data and it relation has been studied. Moreover, in this chapter a detail analysis of data preprocessing, three feature extraction (i.e., TDbFE, FDbFE, and TFDbFE), most relevant feature selection, data visualization, used different software and dataset have been presented and explored. Moreover, a wide range of open access available software and tools and data base have been listed for point of industry and academic uses.

References

[1] Design Thinking. Available from: https://www.ideou.com/pages/design-thinking. Accessed 30.10.2020

[2] Global Big Data Analytics Market, Forecast to 2023. Available from: https://store.frost.com/global-big-data-analytics-market-forecast-to-2023.html. Accessed 30.10.2020

[3] IEA. Smart Grids, IEA, Paris. (2020). Available from: https://www.iea.org/reports/smart-grids.

[4] AWS + C3.ai Application Development Time and Cost Savings, Third-Party Report by AWS Premier System Integrator. (2020). Available from: https://c3.ai/wp-content/uploads/2020/05/C3ai-plus-AWS-accelerates-time-to-value.pdf. Accessed 30.10.2020

[5] NCTA (National Cable & Telecommunications Association) data. Available from: https://www.ncta.com/.

[6] H. Malik, et al., Feature Selection using RapidMiner and Classification through Probabilistic Neural Network for Fault Diagnostics of Power Transformer, In Proc. IEEE Int. Conf. on Emerging Trends and Innovation in Technology (INDICON 2014) (2014), doi: 10.1109/INDICON.2014.7030427.

[7] A. Aggarwal, et al., Feature Extraction Using EMD and Classification through Probabilistic Neural Network for Fault Diagnosis of Transmission Line, In Proc. IEEE ICPEICES-2016, pp. 1-6, 2016. (2016), doi: 10.1109/ICPEICES.2016.7853709.

[8] Y. Pandya, et al., Feature Extraction Using EMD and Classifier through Artificial Neural Networks for Gearbox Fault Diagnosis, Book Chapter in Applications of Artificial Intelligence Techniques in Engineering, Advances in Intelligent Systems and Computing 697 (2018) 309–317, https://doi.org/10.1007/978-981-13-1822-1_28.

[9] S. Mishra, et al., Selection of Most Relevant Input Parameters Using Waikato Environment for Knowledge Analysis for Gene Expression Programming Based Power Transformer Fault Diagnosis, International Journal of Electric Power Components and Systems, 42 (16) (2014) 1849–1862, http://dx.doi.org/10.1080/15325008.2014.956952.

[10] S. Mishra, et al., Selection of Most Relevant Input Parameters Using Principle Component Analysis for Extreme Learning Machine Based Power Transformer Fault Diagnosis Model, International Journal of Electric Power Components and Systems 45 (12) (2017) 1339–1352 https://doi.org/10.1080/15325008.2017.1338794.

[11] A.K. Yadav, et al., Selection of Most Relevant Input Parameters Using WEKA for Artificial Neural Network Based Solar Radiation Prediction Models, Renewable and Sustainable Energy Reviews 31 (2014) 509–519, https://doi.org/10.1016/j.rser.2013.12.008.

[12] S. Syeed, et al., Selection of Most Relevant Input Parameters Using WEKA for Artificial Neural Network Based Concrete Compressive Strength Prediction Model, In Proc. IEEE PIICON-2016 (2016) 1–6, doi: 10.1109/POWERI.2016.8077368.

Chapter 2

Intelligent Data Analytics for PV Fault Diagnosis Using Deep Convolutional Neural Network (ConvNet/CNN)

1 Introduction

In the current scenario, most of the countries of the globe have focused on policies and initiatives to attain sustainability by reducing the conventional energy uses (i.e., fossil fuel based energy) by decreasing the CO_2 emission to meet the Paris agreement objectives [1]. Moreover, service period and reliability of PV module is increasing so as to reduce the capital cost and enhance the utilization. Therefore, the installation and commissioning throughout the world is increasing to meet the power demand by reducing the non-renewable energy sources. So, there is a big challenge for power engineer to maintain the PV panel in healthy condition to meet the consumer demands, such as moderate crystal anomaly striation rings [2].

According to one analysis conducted by D.C. Jorden and coworkers [3], the rate of degradation of PV module (made by crystalline silicon material) is 0.8% per year. Generally, a PV module anomaly is an effect that: (1) degrades module power, or (2) generates a safety problem. While, there are many PV module failures (PVMF), which are excluded by proper categorization such as moderate crystal anomalies, striation rings, light-induced power degradation due to boron-oxygen complex (power reduction by 10%–30%) [4] and Staebler-Wronski effect (SWE). There are also some PVMF, which occur by some external reasons, such as clamping, transport and installation, connection failure, lightning effect/strike, etc. Hence, anomalies of a product are categorized into three main parts: (1) incipient level of anomalies, (2) medium level of anomalies (midlife failure), and (3) wear-out condition. The detailed graphical representation has been shown in Fig. 3.1 (at page 46) of an IEA report [5]. There are different reasons for the anomalies/faults to occur in the PV module as shown in Fig. 2.1 [6], which leads to the power loss of 0%–20%.

Intelligent Data-Analytics for Condition Monitoring. http://dx.doi.org/10.1016/B978-0-323-85510-5.00002-8

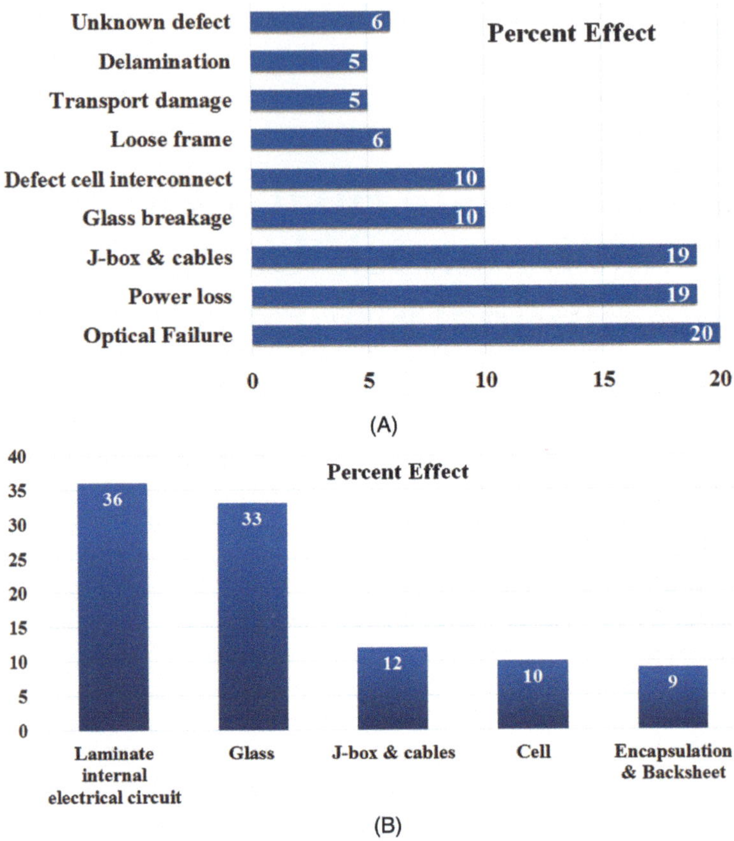

FIGURE 2.1 PV failure rates as per (A) information given by customer within 2 years of installation and (B) information collected by manufacturers within 8 years of installation. *(Data from Ref. [6].)*

There are some conventional methods, which are used to identify the PVMF. The first and most popular method is visual inspection, which is described in IEC61215, IEC61646 Standards. The second popular way is DoVFiF "Documentation of visual failures in the field," which will overcome some of the drawbacks of first method. In summary, the component wise failures types are listed in Table 2.1.

In the 21st century, there are several advance level methods/prototypes available in the market to identify/predict the anomaly/failure condition in PVM. Some of them are: (1) Thermography methods (TM), (2) Electroluminescence (EL) method (ELM), (3) UV-fluorescence (FL) method (UFM), and (4) Single transmission method (STM), etc. The TM are further classified into three categories according to its utilization as listed: (1) TM under steady state conditions (TMuSS), (2) Pulse thermography (PTM), and (3) Lock-in thermography

TABLE 2.1 Type PVMF inspected according to IEC 61215, 61646 Standard.

S. no.	PVMF	Associated component of PVM
1	Bubbles	Front of PVM
2	Delamination	
3	Yellowing	
4	Browning	
5	Broken cell	PV cells
6	Cracked cell	
7	Discolored antireflection	
8	Burned	Cell metallization
9	Oxidized	
10	Bend	Frame
11	Broken	
12	Scratched	
13	Misaligned	
14	Delaminated	Back of module
15	Bubbles	
16	Yellowing	
17	Scratches	
18	Burn	
19	Loose	Junction box
20	Oxidation	
21	Corrosion	
22	Detachment	Wires and connectors
23	Brittle	
24	Exposed electrical parts	

method (LiTM). TM provide the fast response in real-time due to its non-destructive nature for both large as well as small scale PVFM. The PTM requires an external heat source for its operation. So, its process for image capturing is less than few milliseconds, else image becomes blur. LiTM can be used for detection of weak heat sources of both crystalline and thin-film modules. LiTM can be used into two ways as: DLiTM (dark lock-in thermography) and ILiTM (illuminated lock-in thermography), which depends on used sources whether electric current/voltage source or light source, respectively. The ELM is performed in dark environment due to low IR value emitted by PV (~1150 nm).

The UFM can be used for the identification of material degradation. Initially, STM is developed for earth leakage detection. In PVFM, STM can be used to detect the local connections (i.e., open circuit failure).

In this study, an intelligent data analytics for PV fault diagnosis using deep convolutional neural network (ConvNet/CNN) is developed for PVFM. The developed model is validated and demonstrated by using image process approach for real-time implementation.

Organization of this chapter is formulated into seven sections. The introduction part in Section 1 shows different type of PV faults and its associated conventional standard based techniques for diagnosis purpose. The data analysis for the inventions in this area is presented in Section 2, which is useful for the future research planning and for the targeting the market region. The Section 3 shows the details of the experimental study and the whole procedure of dataset collection. The proposed approach framework is represented in Section 4, which includes the following sub-sections: (1) dataset collection, (2) preprocessing, (3) CNN model formulation, and (4) multiple testing of the model. The Section 5 shows the implementation procedure in detail for ConvNet/CNN algorithm which is used for PV fault diagnosis. The results demonstration and its discussion are represented in Section 6 and finally conclusion is shown in Section 7.

2 Intelligent data analysis for photovoltaic module failures (PVMF) analysis

This section includes the intelligent data analytics for PVMF analysis related to its invention. The inventions are analyzed by using different keywords such as "Failures of Photovoltaic Modules" or "fault of Photovoltaic Modules" or "fault detection of Photovoltaic Modules" or "Condition Monitoring of Photovoltaic Modules." This type of data analytics is very useful for the future finding research space for the upcoming new technologies. Therefore, this section includes several valuable information such as profiles of innovation, innovations related to geographic area, available key technologies & methods, analysis of ownership, analysis of involved researchers or inventors and its co-inventors, significant inventions, analysis of market and its importance, innovative word related to the inventions, and analysis of each technology related to PVMF analytics and its associated domain. The section covers the analytics domain in PVMF area from 2001 to 2019 only, which is represented by the applied and granted applications throughout the world as shown in Fig. 2.2.

In general, earlier-mentioned applications are from only two areas, that is, utility side and invention side. In the current scenario, majority of applications are from invention side, that is, around 80% plus. According to their rank, the top 10 countries in this area are China, United States, Korea, Japan, Germany, EPO, France, WIPO, Spain, and India. China covers around 45% plus inventions followed by the United States (around 26% plus). The invention trend from top 10 countries is represented in Fig. 2.3.

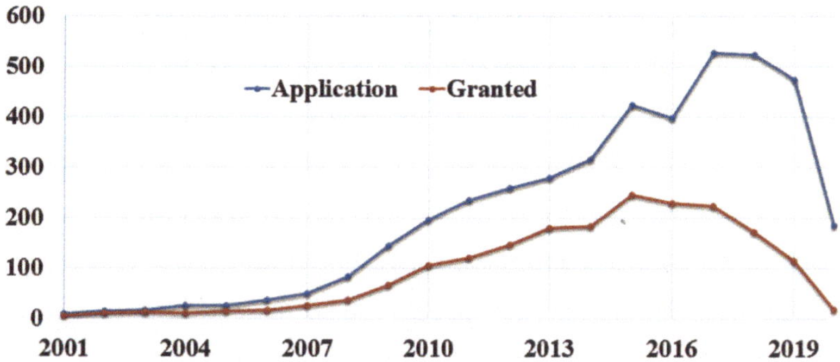

FIGURE 2.2 **Analytics representation of applied and granted applications for PVMF area.**

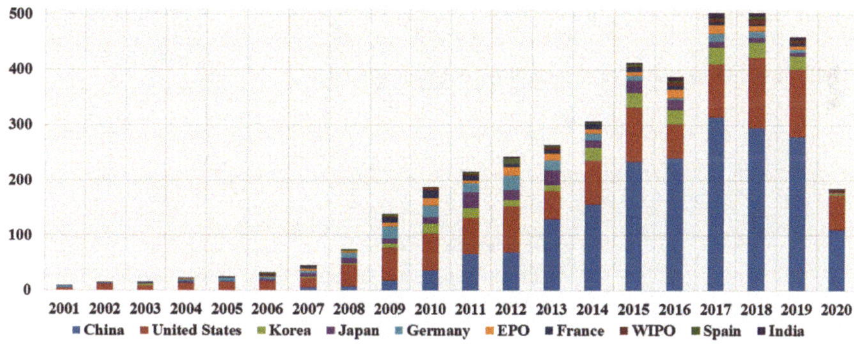

FIGURE 2.3 **Applied application scenario representation for top 10 countries in the area of PVMF analytics.**

In the area of PVFM, the top 10 organization which leads the maximum part of the innovative work in this area are: (1) State Grid Corp Of China, (2) Telefon Ab Lm Ericsson (Publ), (3) Sma Solar Technology, (4) China Electric Power Research Institute, (5) Sunpower Corp, (6) Hohai Univ Changzhou, (7) Sungrow Power Supply, (8) Gree Electric Appliances Of Zhuhai, (9) General Electric Co., and (10) 3M Innovative Properties Co. The detailed information is shown in Fig. 2.4.

Moreover, a specific keyword plays a very important role for proper analysis of innovations available in the world. The innovation word cloud shows the basic idea of the search that which one keyword has more influencing weightage for a specific technology and its associated research innovation. For the PVMF analysis, the word cloud is shown in Fig. 2.5. It is based on earlier-mentioned keyword and it represents the 100 key concepts along with its weightage value based on available inventions. After the analysis of the Fig. 2.5, it is concluded that "photovoltaic module" has highest weightage among all. Moreover, directly related to the PVMF analysis, there are some key concepts such as junction

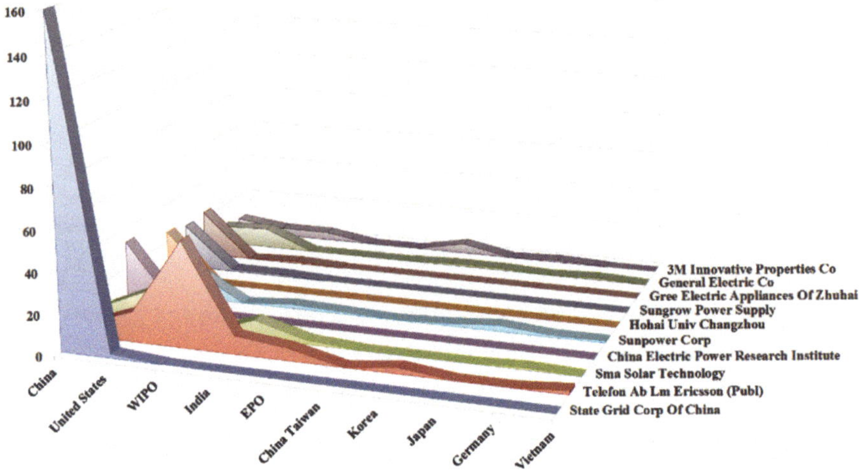

FIGURE 2.4 Geographic representations of top companies.

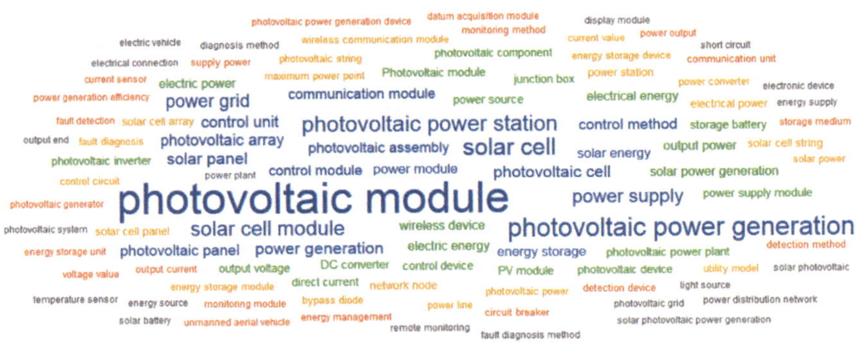

FIGURE 2.5 Word cloud for PVMF analysis.

box (34th rank), fault diagnosis (52nd rank), detection methods (61st rank), monitoring methods (62nd rank), detection devices (63rd rank), monitoring module (69 rank), fault detection (75th rank), current sensor (78 rank), diagnosis method (83 rank), short circuit (91 rank), remote monitoring (93 rank), and fault diagnosis methods (98th rank). The width of the word shows the rank and weightage of the concept in the area of PVMF analysis.

The key take away from this section is that the market value of the invention is very high and China, which plays a role of leader in this domain, is a key territory for the invention and implementation. This type of data analytics is very important for future planning and is a roadmap for the ongoing research with respect to the industry players as well as the launch market prediction of the solution in the area of PVMF analysis.

3 PV image data set collection

In this section, the value of data, experimental design, material, and method have been explained in detail. The experimental dataset for PVMF analysis is collected, which is created by Estefanía Alfaro-Mejía and coworkers [7], and is freely available for research purposes. It includes 277 thermographic images of three different conditions, which are: (1) Snail trails, (2) hot spot failures, and (3) healthy condition. The images are captured by using IR camera of 7–13 μm wavelength. The images are captured during acceptable weather condition such as temperature (26°C–32°C), irradiance (500–1000 W/m^2), and wind speed (3–5 m/s).

Fig. 2.6 represents a schematic of PVM illustration, which shows the panel with 9 × 4 cells in three different conditions such as healthy cell (in white colored), snail trails cell (in orange colored), and hot spot cell (in red colored) failure. The thermal image of this module is captured by using IR thermal camera for further study. For the experimental design point of view, following assumptions are taken care while capturing the images: (1) to measure the weather condition, pyranometer is placed at 3 m height from the panel surface to measure the weather parameters, (2) irradiance level at least 500 W/m^2, (3) image capturing time period between 10:00–11:30 AM and 13:00–14:00 PM, (4) wind speed between 3 and 5 m/s, (5) ambient temperature range 26°C–30°C, and (6) the position of UAV 2 m far from lowest side of panel and 2.3–2.7 m height from panel base. Based on these assumptions, seven experimental sessions are performed to capture 277 images of three conditions of PVM.

The snail tracks defects are normally visible by human eye. Generally, this will occur 3 months to 1 year after the panel installation at site. It depends on the environmental condition as well as different seasons. So, during summer season, it is created in a fast manner. There are several reasons for snail occurrence such

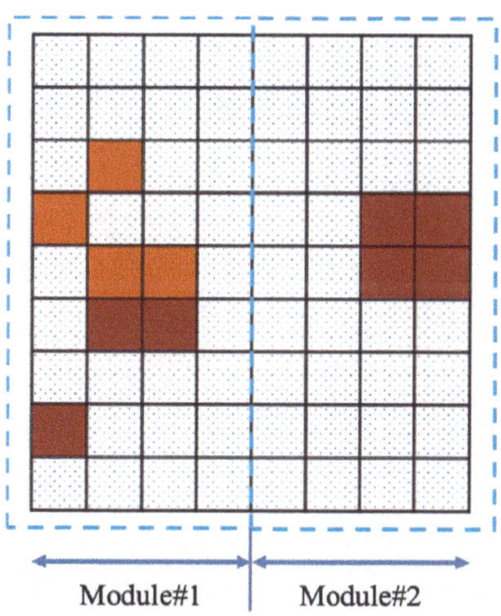

FIGURE 2.6 A schematic of PVM illustration.

as discoloring silver finger and/or normal silver fingers. But discoloring silver finger is highly responsible to snail tracks than normal one. Due to snail tracks, the conductivity of the panel is reduced along the snail lines. Moreover, it is not depended on silver paste of the cell. The high leakage current in the PVM is due to snail tracks. Moreover, the hot spot in the PVM is occurred when bypass diode does not work properly and unable to limit the reverse voltage at the affected area.

4 Proposed approach

Fig. 2.7 shows the proposed approach for PVMF analysis, which comprises in three basic working steps such as: (1) Part-A: data capturing and preprocessing, (2) Part-B: ConvNet/CNN model design, and (3) Part-C: multi-step testing and adaptability. In step#1 of part-A, the formation of experimental setup is created,

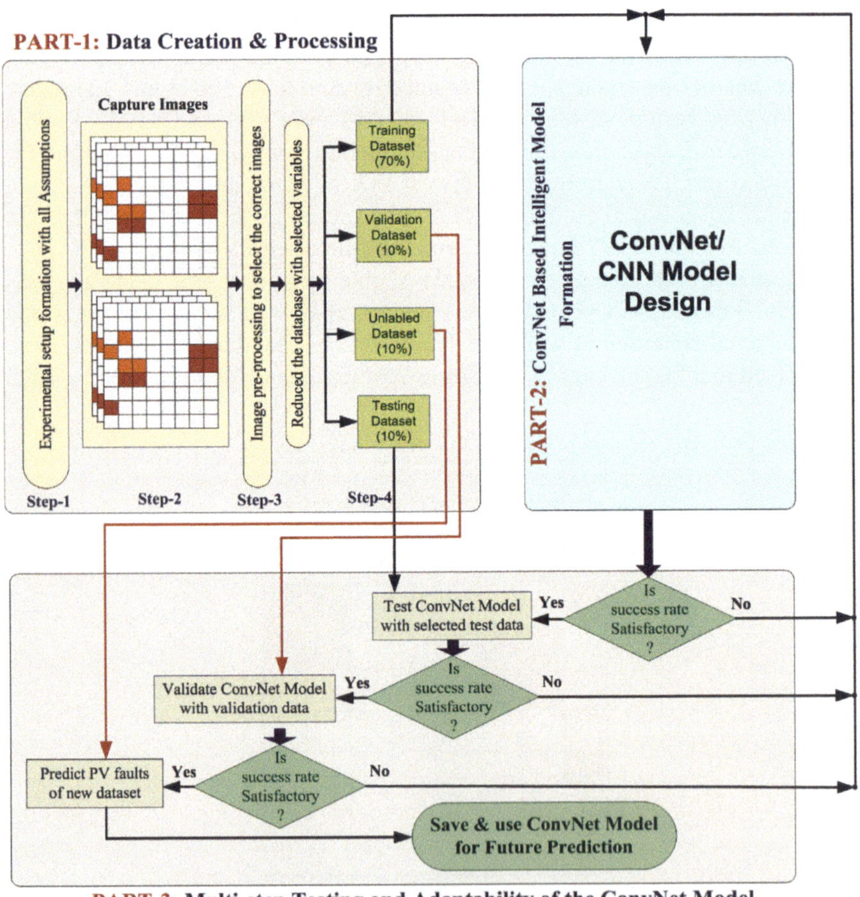

FIGURE 2.7 Proposed approach schematic for PVMF analysis.

which is based on some standard level of assumption so that created images will be good with high resolution. Total 277 images of three different conditions have been created in step#2. The different types of PVM with different defect of cell are presented in different color as shown in Fig. 2.6. In step#3, data pre-processing and image selection is performed. During this process, correct one data has been selected for the further study. In step#4, four different types of data files have been created, which are utilized for training and testing phase of the CNN model. The part-B is related to design the DNN model using ConvNet/CNN algorithm. In part-C of the proposed approach, a multi-step testing process is performed to validate the adaptability of the ConvNet/CNN based DNN model for PVMF analysis and diagnosis.

5 Deep convolutional neural network (ConvNet/CNN)

In the recent research domain, there are several training algorithms for deep neural network (DNN), which are available for implementation in various domains of research. However, now days AlexNet, GoogleNet, VGGNet, and ResNet are found to be used more frequently in most of the applications in the field of innovations/inventions [8]. Google Inception network and ResNet-152 have secured the top rank/position in terms of accuracy and performance in the present list of applications [9]. The evaluation burden of ResNet-152 is very high, that is, 11 billion Floating Point Operations (FLOPS), which are extremely large in comparison to other networks. In contrast, a GoogleNet takes only 2 billion FLOPS with an accuracy of more than 90%, as represented in Fig. 2.8. However, presented schematic is little bit complex due to several parallel layers with different filter sizes. Hence, the AlexNet has been used as the baseline network in this study.

The AlexNet is loaded, and the last three layers are adjusted to improve the suitability of the network according to the PVMF analysis. For the implementation of transfer learning, MATLAB software has been utilized [9]. According to the Fig. 2.8, it is analyzed that the network comprises in a simple formulation (such as 3-stages of max Poolings, 2-stages of dropouts, and 3-fully connected networks), which are as per the standard AlexNet architecture. Fig. 2.8 is little

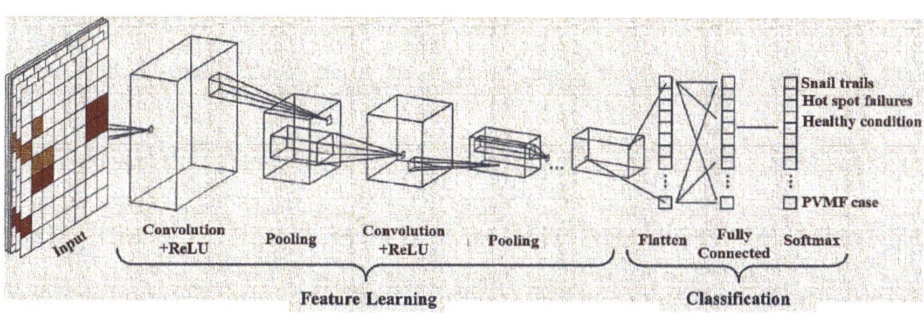

FIGURE. 2.8 A schematic of ConvNet/CNN based deep neural network illustration.

bit different from the standard AlexNet due to 3 more layers (i.e., connected layer, softmax layer, and a classification output) are included to the network for classifying the nonlinear problem.

The input data to this network is an image of the size of 227 × 227 × 3 pixels. According to the Section 3, three-types of PVMF analysis are considered for diagnosis. The images obtained as described in Section 3, are processed with a fixed size of 227 × 227 × 3 to map the input size of the network. A set of images of each type of PVMF (~70%) is prepared as the database for input of network and out of which 30% images are utilized for testing/validation of the model performance. The selected optimal parameters used in training phase are: (1) Weight Learn Rate Factor: 20, (2) Bias Learn Rate Factor: 20, (3) Minimum batch size: 120, (4) Maximum iterations: 1000, (5) Initial Learn Rate: 0.001, (6) Hardware resource: Single CPU, (7) Learning rate schedule: Piecewise, (8) iteration per epoch: 25, and (9) Validation Frequency = 25 iterations.

6 Results and discussion

The ConvNet/CNN is widely utilized tool for DNN. This is generally suitable tool for image classification, although it is utilized for other domain as well. In the implementation of ConvNet, each layer is organized in 3-D input to the 3-D output. For example, The PVMF image input, the first layer of ConvNet (input layer) feed with 3-D input value, which includes the 3-D dimension such as height, width, and the color of the PVMF image then neurons in the first convolution layer joined with region of these PVMF images and then it is transformed into 3-D output. The neurons at each hidden unit learn nonlinear combination of original inputs features, which are known as activation as output of one layer is input to the next layer. Finally, processed features become input to the classifier at the end of the model. Based on this example, the PVMF for different failure condition identification has been implemented and obtained results are tabulated in Table 2.2 and its learning curve is represented in the Fig. 2.9.

TABLE 2.2 ConvNet/CNN based performance analysis for PVMF diagnosis model.

	MSE	RMSE	Accuracy (%)	CPU processing speed (s)	GPU processing speed (s)
Training phase	7.95	2.82	92.05	23.215	2.59
Testing phase	16.5	4.06	83.5	1.45	0.012

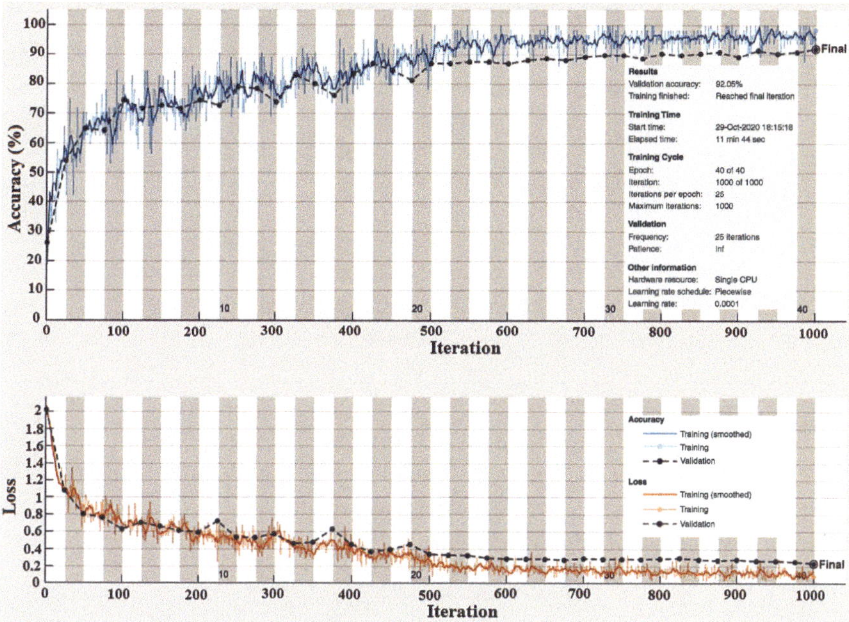

FIGURE 2.9 Performance plot for PVMF analysis.

Fig. 2.10 shows the heat map for ConvNet, which shows how strongly each hidden unit activates and highlights how the activations change over time. Fig. 2.11 represents the each case of PVM case and its corresponding accuracy. The overall accuracy for ConvNet model is 92.05% and 83.5% for training and testing phase, respectively. The main key takeaway form this study is to

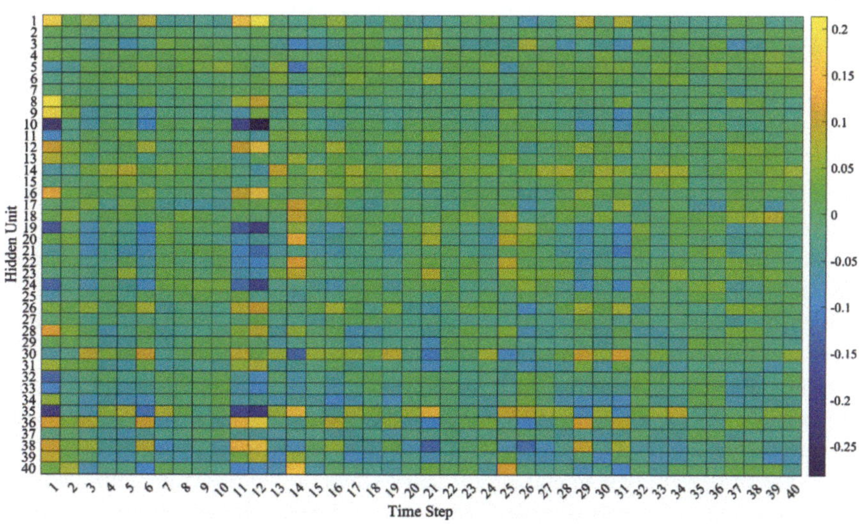

FIGURE 2.10 ConvNet/CNN activation for selected hidden units for PVMF analysis model.

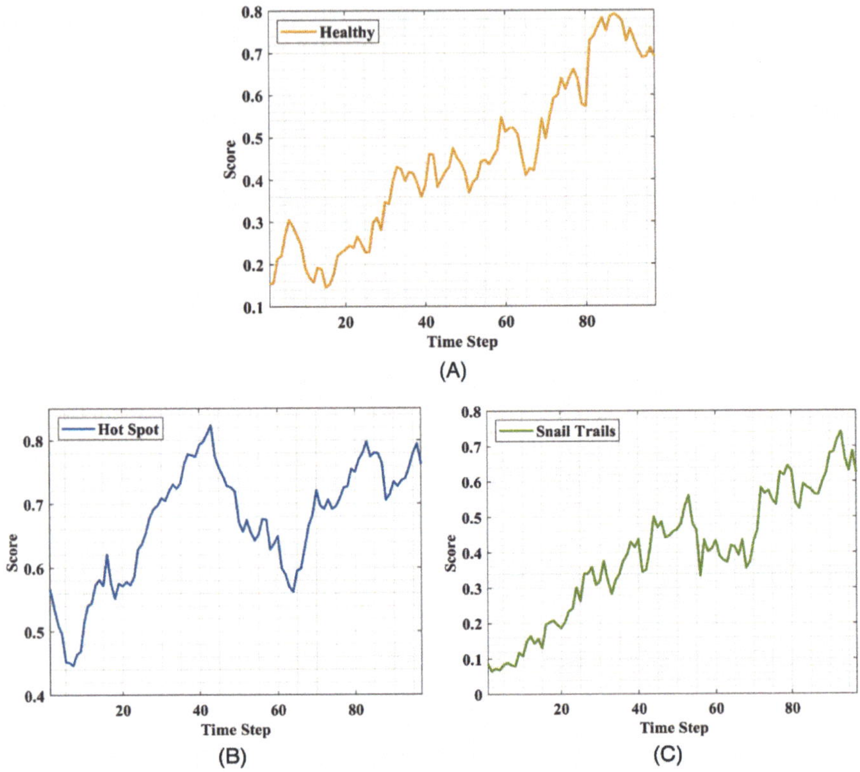

FIGURE 2.11 Performance evaluation representation for each condition of PVM as (A) Healthy PVM, (B) Hot spot PVM, and (C) Snail trails PVM case accuracy.

know the step-wise-step procedure to implement the ConvNet model for PVMF analysis. The prediction can be enhanced by performing same study in different way. This research gap may be covered in future prospective studies.

7 Conclusion

In this study, a new DNN approach based on ConvNet algorithm has been implemented for the PVM failure analysis using high resolution image dataset of 277 images. In this chapter, most of all possible failure cases of PVM system have been covered and its associated conventional techniques were also studied to identify the research gap. The experimental dataset has been collected and preprocessed for the further application. Developed DNN approach based on ConvNet algorithm is trained and tested multiple times to identify the adoptability scenario so that it can be used on real side of the PVM. The obtained results are outperformed, which shows the adaptability of the proposed approach. The main idea of this study is to make familiar of DNN model implementation based on ConvNet algorithm in the area of PVM failure analysis which has not been reported in the literature yet.

For the future scope of the work, the different feature extraction, feature selection, and diagnosis algorithm may be implemented to diagnose the PV fault with high diagnosis accuracy.

References

[1] J. Nieto, O. Carpintero, L. Miguel, Less than 2C? An economic-environmental evaluation of the Paris Agreement, Ecol. Econ. 146 (2018) 69–84 doi: https://doi.org/10.1016/j.ecolecon.2017.10.007.

[2] M.W. Ahmad, et al., A Fault Diagnosis and Postfault Reconfiguration Scheme for Interleaved Boost Converter in PV-Based System, IEEE Transactions on Power Electronics 36 (4) (2021) 3769–3780, doi: 10.1109/TPEL.2020.3018540.

[3] D.C. Jordan, S.R. Kurtz, Photovoltaic degradation rates—an analytical review, Prog. Photovolt Res. Appl. 21 (12–29) (2011), doi: 10.1002/pip.1182.

[4] A. Shah, W. Beyer, Thin-film Silicon Solar Cells, In A. Shah (Ed.), EPFL Press, 2010, pp. 30–35.

[5] M. Kontges, S. Kurtz, J. Ulrike, K.A. Berger, K. Kato, T. Friesen, in: Review of Failures of Photovoltaic Modules, vol. 1911, 2015, 2016. [Report IEA-PVPS T13-01:2014]. Available from: https://repository.supsi.ch/9645/1/IEA-PVPS_T13-01_2014_Review_of_Failures_of_Photovoltaic_Modules_Final.pdf. Accessed 28.10.2020.

[6] D. DeGraaff, R. Lacerda, Z. Campeau, Degradation Mechanisms in Si Module Technologies Observed in the Field; Their Analysis and Statistics, Presentation at PV Module Reliability Workshop, NREL, Denver, Golden, USA, 2011. Available from: http://www1.eere.energy.gov/solar/pdfs/pvmrw2011_01_plen_degraaff.pdf.

[7] E. Alfaro Mejia, H. Loaiza Correa, É. Franco-Mejia, A.D. Restrepo-Girón, S. Nope, Photovoltaic system thermal images, Mendeley Data 2 (2019). doi: http://dx.doi.org/10.17632/82vzccxb6y.2.

[8] G. Sapijaszko, W.B. Mikhael, An overview of recent convolutional neural network algorithms for image recognition, in: 2018 IEEE Sixty-First International Midwest Symposium on Circuits and Systems (MWSCAS), Windsor, Ontario, Canada, 2018, pp. 743–746. doi:10.1109/MWSCAS.2018.8623911.

[9] Difference between AlexNet, VGGNet, ResNet, and Inception. Available from: https://towardsdatascience.com/the-w3h-of-alexnet-vggnet-resnet-and-inception-7baaaecccc96#:~:text=AlexNet%20and%20ResNet%2D152%2C%20both,training%20time%20and%20energy%20required. Accessed 28.10.2020.

Chapter 3

Intelligent Data Analytics for Power Transformer Health Monitoring Using Modified Fuzzy Q Learning (MFQL)

1 Introduction

The stability of wind energy conversion system (WECS) depends on the reliability of its sub-systems and all associated components. Wind turbine power transformer (WTPT) is a vital component among all components used in WECS, also a key component in an electrical power network (i.e., transmission line network and/or distribution line network). If any catastrophic failure condition happens in the working scenario of WTPT, it will decrease the performance as well as stability of WECS or power system or may lead to a complete blackout of the area. Malfunctioning of a WTPT is an economic loss, which results in faults such as electrical discharge fault and thermal fault. These faults ought to be monitored for correction and proper functioning of the power system. According to the one statistics of the failure rates for the sub-assembly of WECS, the power transformer in electrical power network (PTiEPN) leads upto 9.92% failures, which is the fourth highest failure rate in ranking after the tower (12.38%), gearbox (10.98%), and rotor blades (10.16%) as shown in the Fig. 3.1 [1].

The root causes of transformer failure are mechanical stress, electrical stress, and thermal stress and these root causes are further classified into two major groups, which are: (1) internal failures (loss of winding, overheating, solid contaminations in oil, oxygen, partial discharge, moisture, and design defects, etc.) and (2) external failures (i.e., lightening, insulation deterioration, system failure, winding resonance, system overload, system switching, short circuited condition, etc.) as shown in Fig. 3.2. These root causes are generally generated in the bushing, auxiliaries, tap-changer, and main tank [1]. Most of available transformer failures are classified into anomaly at incipient levels,

Intelligent Data-Analytics for Condition Monitoring. http://dx.doi.org/10.1016/B978-0-323-85510-5.00003-X

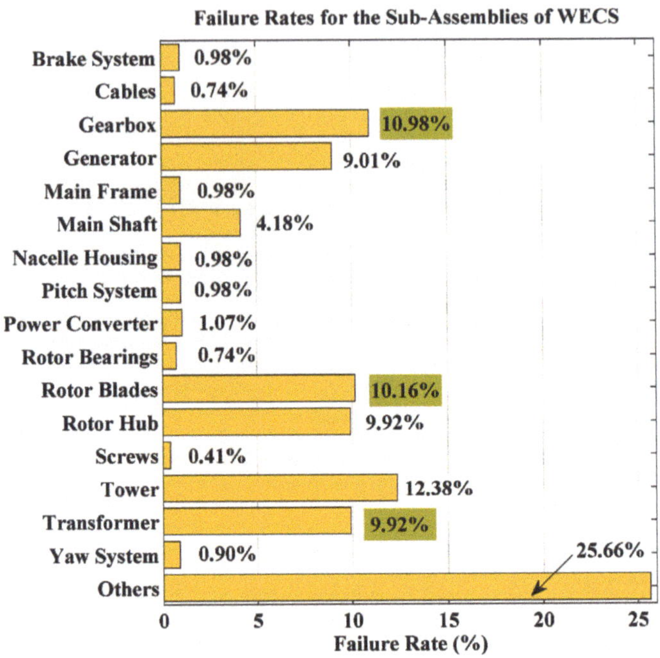

FIGURE 3.1 **Failure rates for sub-assemblies of WECS.**

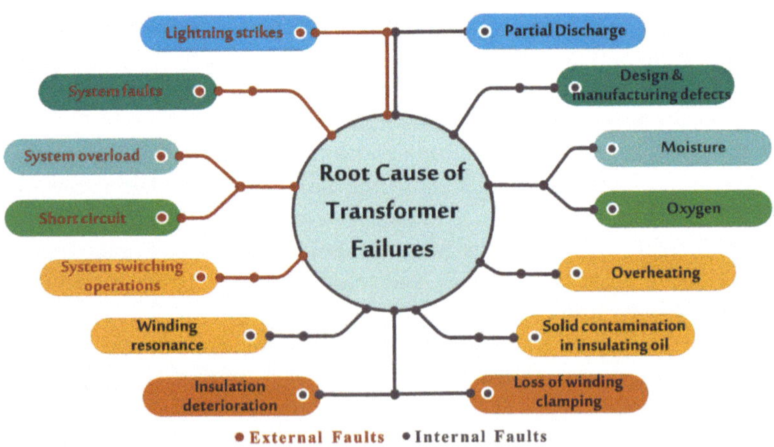

FIGURE 3.2 **Root causes of transformer failures.**

which are further categorized into four sub-categories [1], and are represented in Fig. 3.3 for easy understanding.

Generally, nine gases (G1:CH_4, G2:C_2H_6, G3:C_2H_4, G4:C_2H_2, G5:H_2, G6:CO_2, G7:CO, G8:O_2, and G9:N_2) are generated inside the transformer due to fault (i.e., PDF: partial discharge fault, AF: arcing fault, COF: cellulose

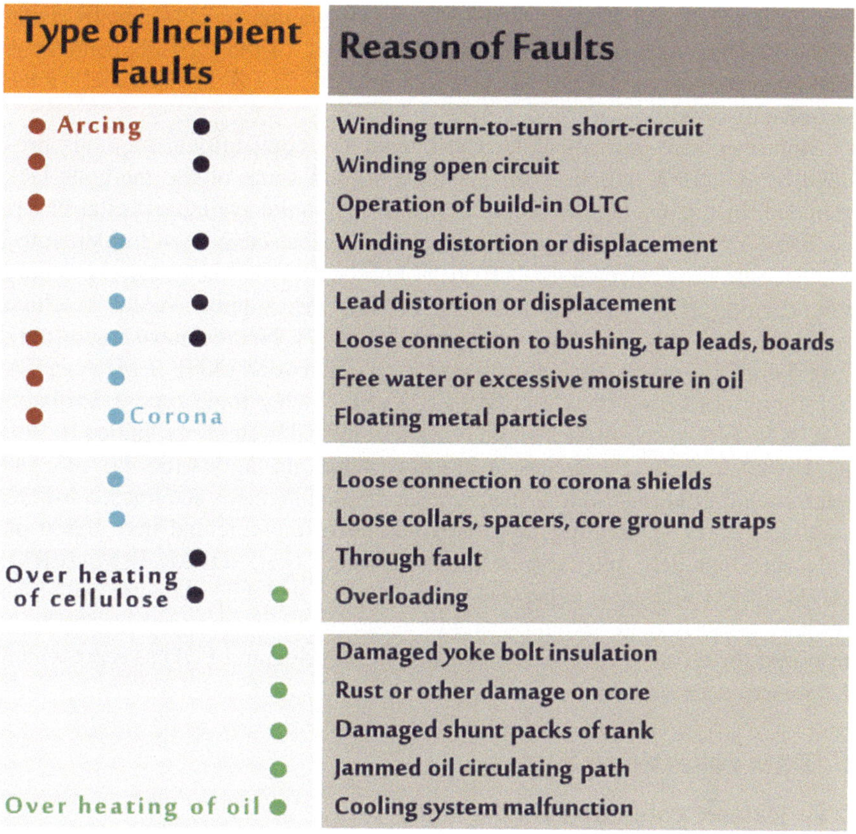

FIGURE 3.3 **Reasons of power transformer anomaly at incipient level.**

overheating fault, OOF: oil overheating fault, TDF: thermal decomposition fault, TF: thermal fault, HEDF: high energy discharge fault). Some of them are dissolved in the oil (i.e., G1–G6) and some are not dissolved in the oil (G7–G9) [2,3]. These generated fault gases are also categorized into hydrogen group (G1–G6) and carbon oxide group (G7 and G8). As these generated gases are due to occurrence of faults in the transformer, so these gases are used further for the incipient fault detection and diagnosis (FDD) proposes. To do so, a dissolve gas analysis (DGA) is performed by utilizing the ppm level of each gas. These ppm levels of the gases are evaluated by using DGA analyzer (such as Kalmen Transport X, United States). Then IEEE and/or IEC methods (i.e., IDRM: IEEE Doernenburg ratios method, IERM: IEC ratios method, IKGM: IEEE Key Gas method, IRM: IEEE Roger ratios method and IDTM: IEC Duval triangle method) are implemented to identify the FDD level of the PTiEPN. The IKGM can detect the four different types of faults (i.e., COF, AF, OOF, and PDF) by using six DGA key gases (G1–G5 and G7). Similarly, IDRM can detect three faults conditions (AF, PDF, and TDF) of the PTiEPN. The IRM can detect the five types (PD, TF, HEDF) of condition of the PTiEPN. Similarly, IERM can analyze the HEDF, TF, D1, D2, and PDF. The IDTM can analyze the six anomaly conditions of the PTiEPN.

Moreover, the anomaly detection rate of the conventional methods provided by the IEEE and IEC are not upright, and some of the methods lack from healthy condition identification as well [4]. In the available digital libraries, there are several intelligent techniques (IT) that have been implemented to overcome the drawback of IEEE/IEC approaches such as gene expression programming (GEP) [4], Immune-NN (INN) [5], support vector machine (SVM) [6], particle swarm optimization (PSO) [7], Neuro-Fuzzy logic [8,9], Levenberg-Marquardt algorithm based neural network (NN) (LMNN) [10], and many more (i.e., Proximal-SVM (PSVM), Fuzzy logic type-2, Probabilistic-NN (PNN), cognitive system (CS), and hybrid system) explained in Ref. [4]. However, such approaches of ITs have some limitations too. So, to overcome such limitation and to achieve the desirable diagnosis accuracy, a MFQL (modified fuzzy Q learning) approach has been implemented here based on eight different data set used as input variables (i.e., IDRM, IRM, IERM, IKGM, IDTM variables, generated vector H of 24 variables, selected variables by PCA, and selected variables by J48 algorithm) and an approach has been proposed for PTiEPN of WECS condition monitoring and FDD as represented in sub-sequence sections.

2 Data collection/source

2.1 Dataset collection for the study

Total 552 DGA data samples are utilized in this study [4]. These data samples are collected from the credible literature (192 data samples) and emulated practical data in the laboratory (360 data samples). All collected data samples include eight different type of faults condition along with healthy scenario of the transformer. The representation of these fault conditions are: (1) HC = healthy condition, (2) TFC = Thermal Fault, (3) $z1$ = TFC (<150°C), (4) $z2$ = TFC at (150°C–300°C), (5) $z3$= TFC at (300°C–700°C), (6) $z4$= TFC at (>700°C), (7) $z5$ = PDF of low energy (PD1), (8) $z6$= PD of high energy (PD2), (9) $z7$= LE discharge (D1), (10) $z8$= HE discharge (D2).

The number of data samples for each case are as follow: 31 data samples for HC, 83 data samples for z1-z3, 85 data samples are for z4, 39 data samples are for z5-z6, 130 data samples are for z7, and finally z8 has 184 data samples.

3 Proposed approach and methodologies

The proposed approach for the FDD of a PTiEPN has been represented in Fig. 3.4, which includes two parts. The PART-1 is related to the data set generation and preprocessing the generated dataset. Here, oil samples are

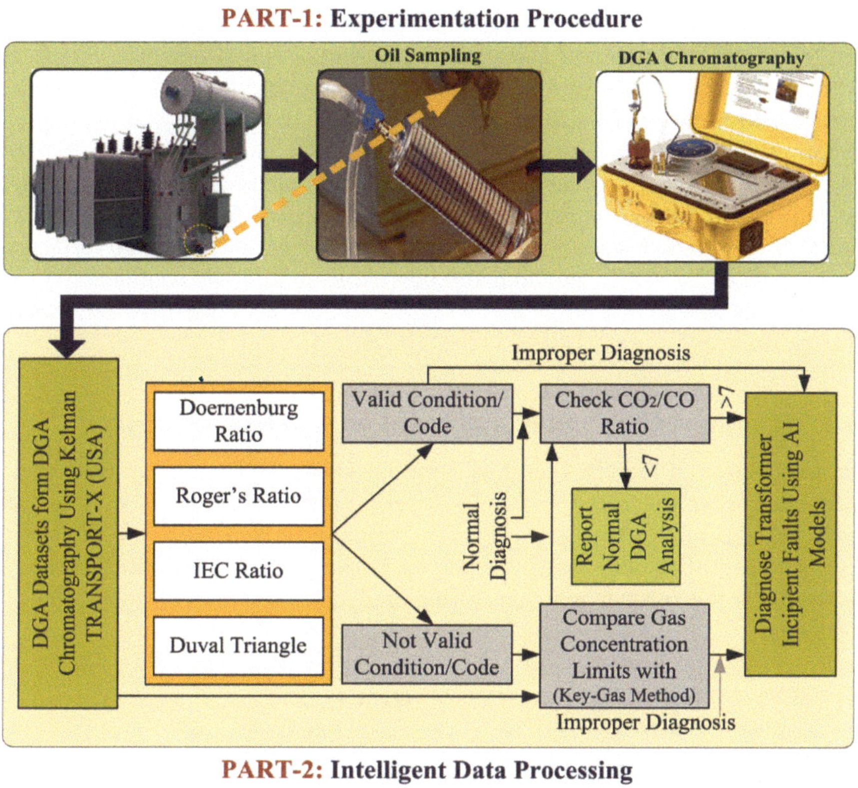

FIGURE 3.4 **Methodology and strategies for implementation of transformer fault diagnosis model.**

collected from the PTiEPN site in an oil cell, which is tested in the laboratory by using DGA chromatography analyzer that is called GDA analyzer. Here, Kalman Transport-X from the United States is used for performing the chromatography of the oil samples. Then collected ppm level of each gas of each sample is utilized for further procedure in the PART-2 of the proposed approach.

The part-2 of the proposed approach is related to intelligent data processing, which belongs to the intelligent data analytics using AI/machine learning approaches. Here in this part, conventional methods of IEC/IEEE are applied first to check the interpretation capability based on the collected 552 data samples. Thereafter, AI/Machine learning approaches are implemented for further analysis. In this study, MFQL is designed and implemented first then its performance for anomaly detection of PTiEPN is compared with the ANN method. The MLP type of ANN method is designed here to validate the proposed approach accuracy.

In this chapter, two types of study have been performed as: (1) without selection of most relevant variables and (2) with selection of relevant variables. The procedure of the most relevant input variables selection is known as feature selection. So, in this study, features are selected by using two different feature selection approaches that are J48 algorithm and PCA method. Moreover, before going to implement these two types of analysis (with and without selection of feature), the performance of the proposed MFQL approach and MLP-ANN method have been analyzed with respect to the input variables coming from IEC/IEEE methods directly. After getting all confidence level of the proposed approach, it can be suggested for the online real-site implementation.

3.1 Feature vector formulation based on standard techniques

In this section, a feature vector of 24 input variables have been generated, which includes the 7 key gases, 10 different ratio values derived from the combination of the key gases and 7% values (derived from the different combination of the key gases) as shown in Eq. (3.1). The feature vector H is used for the further analysis in this study as explained in subsequence sections.

$$H = \begin{bmatrix} C_2H_2, H_2, C_2H_4, CH_4, CO_2, C_2H_6, CO, X, Y, Z, P, Q, R, S \\ \left(\frac{\mathrm{CH_4}}{\mathrm{H_2}}\right), \left(\frac{\mathrm{C_2H_2}}{\mathrm{C_2H_4}}\right), \left(\frac{\mathrm{C_2H_4}}{\mathrm{C_2H_6}}\right), \left(\frac{\mathrm{C_2H_2}}{\mathrm{CH_4}}\right), \left(\frac{\mathrm{C_2H_6}}{\mathrm{C_2H_2}}\right), \left(\frac{\mathrm{CO_2}}{\mathrm{CO}}\right), \\ \left(\frac{\mathrm{C_2H_2}}{\mathrm{H_2}}\right), \left(\frac{\mathrm{C_2H_4}}{\mathrm{CH_4}}\right), \left(\frac{\mathrm{CO}}{\mathrm{CO_2}}\right), \left(\frac{\mathrm{C_2H_6}}{\mathrm{CH_4}}\right) \end{bmatrix}_{552\times 24} \tag{3.1}$$

Where, $\mathrm{X = \%CH_4 = \frac{100*x}{(x+y+z)}}$, $\mathrm{Y = \%C_2H_2 = \frac{100*y}{(x+y+z)}}$, $\mathrm{Z = \%C_2H_4 = \frac{100*z}{(x+y+z)}}$,

$\mathrm{P = \%H_2 = \frac{100*H_2}{(H_2+C_2H_6+CO+CO_2)}}$, $\mathrm{Q = \%C_2H_6 = \frac{100*C_2H_6}{(C_2H_6+m+n+o)}}$,

$\mathrm{R = \%CO = \frac{100*CO}{(C_2H_6+m+n+o+CO+CO_2)}}$, $\mathrm{S = \%CO_2 = \frac{100*CO_2}{(C_2H_6+m+n+o+CO+CO_2)}}$

3.2 Most influencing features selection

The most relevant input variable section is a process to reduce the redundant features/attributes from the feature vector so that FDD accuracy of the classifier may achieve as per satisfactory level as well as the computational burden can be reduced for the point of online implementation. In this chapter, two different approaches for feature selection have been implemented to identify the best one solution for FDD of PTiEPN. These approaches are: (1) J48 algorithm and (2) PCA algorithm. The detailed explanation and mathematical implementation for J48 algorithm and PCA algorithm are represented in Refs. [11] and [12], respectively.

In this study, J48 algorithm is implemented first by using WEKA based platform and obtained results are represented in Table 3.1, which show the rank value corresponding to each input variable as mentioned in Eq. (3.1). Moreover, the all other performance parameters of the J48 algorithm have been represented in Tables 3.2–3.4 and its graphical representation of the decision tree is shown Fig. 3.5. The nodes of the Fig. 3.5 represent the selected attributes. The main beautify of the J48 algorithm is that it remove all redundant variables directly and presents the most relevant variables in form of its node. Therefore, after the analysis of the Tables 3.1–3.4 and Fig. 3.5, it is concluded that J48 algorithm select only eight most relevant variables for the proposed approach. The selected variables are: (1) $\%C_2H_2$, (2) $\%C_2H_4$, (3) C_2H_4/C_2H_6, (4) $\%CH_4$, (5) C_2H_6, (6) CH_4, (7) CO_2, and (8) C_2H_2.

After the J48 algorithm implementation, now, PCA algorithm is implemented based on RapiMiner platform to select the most relevant attribute by using input data matrix of Eq. (3.1). The main problem in PCA algorithm is that it generates a new rank for variable during each run of the algorithm. So, user need to check accuracy level after each run of PCA and then need to remove least ranked variable from the data matrix. Hence, PCA is a lengthy process in comparison of J48 algorithm. Again eight variables have been selected by the PCA algorithm similar to J48 algorithm but selected variables are different as shown here: (1) $\%C_2H_2$, (2) $\%CH_4$, (3) $\%C_2H_6$, (4) $\%H_2$, (5) C_2H_2/CH_4, (6) C_2H_2/H_2, (7) C_2H_4/C_2H_6, and (8) CH_4/H_2.

TABLE 3.1 Importance factor (IF) inform of rank for input variables (IV).

IV	$\%C_2H_2$	C_2H_4/C_2H_6	$\%C_2H_4$	$\%CH_4$	C_2H_6	CH_4
IF	0.928	0.851	0.808	0.80	0.733	0.694
IV	CO_2	C_2H_2	C_2H_4/CH_4	CH_4/H_2	H_2	C_2H_4
IF	0.671	0.568	0.541	0.441	0.418	0.41
IV	C_2H_2/C_2H_4	C_2H_6/CH_4	CO_2/CO	CO/CO_2	C_2H_2/CH_4	$\%H_2$
IF	0.382	0.38	0.285	0.254	0.215	0.198
IV	C_2H_6/C_2H_2	pC_2H_6	$\%CO$	C_2H_2/H_2	CO	$\%CO_2$
IF	0.173	0.162	0.116	0.112	0.101	0.098

TABLE 3.2 Performance analysis of DGA interpretation selected most relevant input variables by J48 algorithm.

Used data matrix	All attribute Eq. (3.1)	Selected feature	FDD accuracy (%)		J48 algorithm performance						Size of model	
			Accuracy	Misclas-sification	Kappa statistic	MAE	RMSE	RAE	RRSE	Computation burden (second)	No. of leaves	Size of the tree
DGA data [326 × 24]	24	8	98.466	1.5337	0.981	0.0465	0.0987	16.916	26.651	0.01	8	17

TABLE 3.3 Class wise detailed performance analysis J48 model.

	TP rate	FP rate	Precision	Recall	F-measure	MCC	ROC area	PRC area	Class
	0.986	0.000	1.000	0.986	0.993	0.991	1.000	1.000	NF
	1.000	0.000	1.000	1.000	1.000	1.000	1.000	1.000	PD
	0.923	0.004	0.980	0.923	0.950	0.942	0.999	0.995	D1
	1.000	0.012	0.959	1.000	0.979	0.0.974	1.000	0.999	D2
	1.000	0.004	0.981	1.000	0.991	0.989	1.000	1.000	T1T2
	1.000	0.000	1.000	1.000	1.000	1.000	1.000	1.000	T3
Weighted avg.	0.985	0.004	0.985	0.985	0.985	0.981	1.000	0.999	

TABLE 3.4 Class wise confusion matrix for DGA interpretation.

HC	DP	D1	D2	T1T2	T35	<- classified as
68	0	1	0	0	0	NF
0	32	0	0	0	0	PD
0	0	48	3	1	0	D1
0	0	0	71	0	0	D2
0	0	0	0	53	0	T1T2
0	0	0	0	0	49	T3

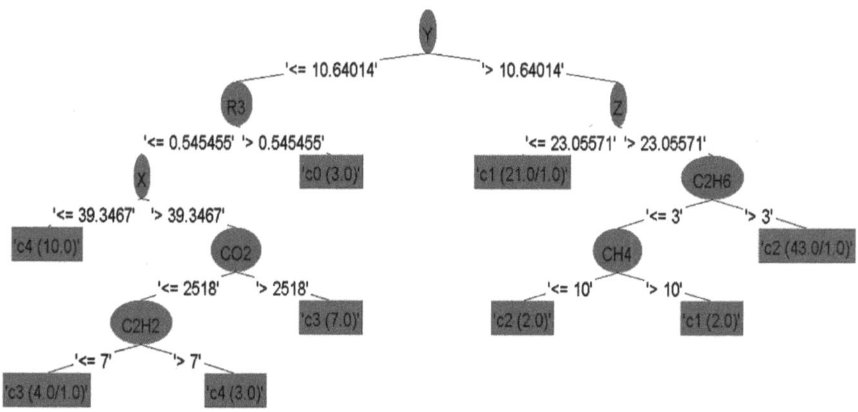

FIGURE 3.5 **Decision tree of selected 8 input variables from 24.**

3.3 Transformer health monitoring techniques

3.3.1 Modify fuzzy Q learning (MFQL) based DGA interpretation

The MFQL implementation is based on the unification of the Q learning, and fuzzy Q learning. Thereafter, MFQL is the modified version of FQL, which is represented in Fig. 3.6. The implementation of MFQL is comprised into 10 steps starting from loading the input data [Eq. (3.1)] till performance evaluation. In the implementation, first input data matrix is mapped as per the fuzzy membership function then these membership functions are used as the input to the classifier. For more mathematical detail of the MFQL implementation for classification problem, reader may refer Ref. [13]. For the fair comparison of the MFQL, MLP based ANN model is implemented based on BP algorithm which is presented in subsequence section.

PART-1: Experimentation Procedure

Measured ppm value from DGA Chromatography → Feature Extraction Using IEC/IEEE DGA Interpretation Method → Feature Selection

PART-2: Data Processing

Step1)	Match Input IMF's to their corresponding labels in each rule to generate a rule firing strength vector $\alpha_i\ \forall i \in N$ as per Eq. of membership function $\alpha_{IMF}(s_1)$	$\alpha_{IMF}(s_1) = e^{-(s_1-c_1)^2/2\sigma_1^2}$ where, c_1= central value and σ_1= STD= $\frac{(s_{max}-s_{min})}{5}$, and range of s=2-100
Step2)	Find the action with the highest q value y_i^* as per Eq. under each rule	Eq.: $y_i^* = \arg\max_{y\in Y} q(i,y)\quad \forall i \in N$, where q value is $q^*(i,y) = \max_{y\in Y} q(i,y)$
Step3)	Compute aggregate action specific firing strength values $\alpha(y^*)\forall y \in Y$ as per Eq.	Eq.: $\alpha(y^*) = \alpha(y_i^*) + \alpha(i, y_i^*)\ \forall i \in N\ \forall y^* \in Y$ Where, $\forall y^* \in Y$ is a 1×6 vector for 6 conditions
Step4)	Implement cooperation among fuzzy rules that have been fired, i.e., $\alpha_i > 0$ by aggregating outputs of each rule to generate a global target value $V(s^k)$ as per Eq. Where T_x^i = linguistic term for variable s_x^k under rule R_i, its membership function is denoted by $\alpha_{T_x^i}$	Eq.: R_i: If s_1^k is T_1^i and.......and s_n^k is T_n^i then $y = y_1$ with $q(i,1)$ or $y = y_2$ with $q(i,2)$ .... or $y = y_m$ with $q(i,m)$
Step5)	Identify the action/fault with the highest reward/q value as the one with the maximum cumulative firing strength as per Eq.	Eq.: $y_{Iden}(s^k) = \arg\max_{y\in Y} \alpha(y)$
Step6)	Generate reward/punishment value for fault identified by the MFQL Classifier as per the reward function *(r)*	$r = \begin{cases} +10 & \text{if} \quad y_{Iden} = y_{true} \\ -10 & \text{if} \quad y_{Iden} \neq y_{true} \end{cases}$
Step7)	Compute temporal difference (TD) error for aligning the q values as per Eq.	Eq.: $\Delta Q = r + \gamma V(s^{k+1}) - Q(s^k, y(s^k))$
Step8)	Update rule specific $q(i,y)$ values as per Eq. (1) and the eligibility traces $e(i,y)$ as per Eq.	$e(i,y) \leftarrow e(i,y)\times\gamma\times\xi + \frac{\alpha_i}{\sum \alpha_i}$ for $y = y_{Iden}$ $\leftarrow e(i,y)\times\gamma\times\xi$ *for* $y \neq y_{Iden}$
Step9)	Reduce exploration rate as per the schedule to explore more in the beginning and less when experience has been gained as per Eq.	Eq.: $\in \rightarrow \frac{\in}{2}$ *every 50 samples* till $\varepsilon = 0.01$
Step10)	Repeat step1 to step 9 till q values converge, i.e., $\Delta q \rightarrow 0$ or becomes a small enough value. This is also manifested in the form a rising classification accuracy	$SuccessRate = \frac{success}{success + failure}\times 100\%$

PART-3: MFQL Based Intelligent Model Formation

Save the Model for Transformer Fault Diagnosis

FIGURE 3.6 **Proposed approach implementation steps.**

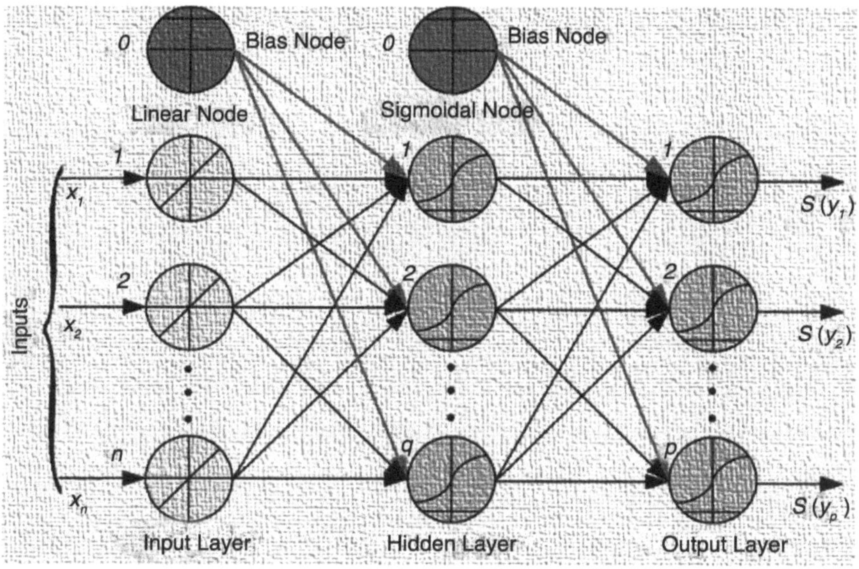

FIGURE 3.7 **ANN architecture for RUL evaluation.**

3.3.2 Multilayer perceptron neural network (MLP-NN)

ANN with one hidden layer is designed as per the formulated mathematical model mentioned in Eqs. (3.2)–(3.8) which includes weight value $W^I = [w^I_{ih}]_{(n+1)\times q}$ and $W^H = [w^H_{hj}]_{(q+1)\times p}$ in between input-and-hidden layer, and between hidden-and-output layer respectively. Fig. 3.7 encapsulates the whole simulation of MLP-ANN architecture [14]. For further reading regarding ANN applicaions, reader may refere [15–19].

At input Layer: the computation at the i_{th} neuron is:

$$\text{Data sample selection} = S(x_i) = x_i,\ i = 1, 2, ..., n; \tag{3.2}$$

and

$$S(x_0) = 1. \tag{3.3}$$

At hidden Layer: The computation at the h_{th} neuron is:

$$\text{Input value } net\,(Hi) = z_h = \sum_{i=0}^{n} w^I_{ih}\, S\,(x_i),\ h = 1, 2, ..., q \tag{3.4}$$

$$\text{Output value } out\,(Hi) = S\,(z_h) = \frac{1}{1+e^{-z_h}},\ h = 1, 2, ..., q \tag{3.5}$$

$$S(z_0) = 1. \tag{3.6}$$

At output Layer: The computation at the j_{th} neuron is:

$$\text{Input value } out\,(Oi) = y_j = \sum_{h=0}^{q} w_{hj}^{H}\, S(z_h),\ j = 1,\ 2, \ldots, p \tag{3.7}$$

$$\text{Output value } out\,(Oi) = S\left(y_j\right) = \frac{1}{1+e^{-y_j}},\ j = 1,\ 2, \ldots, \text{p}. \tag{3.8}$$

where, $S(\cdot)$ = activation function.

4 Diagnosis performance analysis of standard techniques

4.1 Performance analysis of standard techniques without AI

The DGA interpretation of IEEE and IEC conventional methods are performed to validate the minimum and maximum accuracy level of each method (i.e., IFRM, IERM, IKGM, IRM, and IDTM). Based on 349 data samples of laboratory, the interpretation results are presented in Table 3.5 and Fig. 3.8. After analyzing the Table 3.5 and Fig. 3.8, it is concluded that IDTM has high diagnosis accuracy of 70.03% and lowest diagnosis accuracy of 49% with IKGM model.

4.2 Performance analysis of standard techniques with AI

Two different types of artificial intelligence (AI) techniques (i.e., MLP-ANN [14] and MFQL [13]) have been utilized for comparative study of their classification capability in transformer incipient fault diagnosis based on DGA data. The detailed step-wise procedure for implementation of these techniques has been presented in Sections 3.3.1 and 3.3.2. Brief detail of each technique along with its fault diagnostic model has been presented. The results of the proposed diagnostic models for fault diagnosis by using MLP-ANN, and MFQL are represented in subsequence sections, respectively. The diagnostic models using MLP-ANN, and MFQL are the multi-class model and it can classify all conditions in a single instant.

4.2.1 MLP-ANN based power transformer fault diagnosis

Based on conventional DGA interpretation methods, five different models have been implemented by using multi-class MLP neural network as listed in Table 3.6. After comparing Table 3.6 with Table 3.5, it has been found that MLP based DGA interpretation accuracy has been enhanced as compared to conventional DGA interpretation accuracy. The highest diagnosis accuracy to be achieved by MLP model#4 (based on Duval Triangle) is 93.6% (for training phase) and 92.2% (for testing phase). But overall highest diagnosis accuracy is not desirable. Therefore a proposed method is developed in this study to solve the problem of lower diagnosis accuracy.

TABLE 3.5 Performance analysis of IEC/IEEE interpretation method.

Method	Fault	Prediction number	Successful prediction	Successful percentage	Average accuracy
IDRM	TF	113	101	89.38%	64.62%
	DF	152	114	75%	
	NF	17	16	94.11%	
	OC	67	0	0%	
IRM	TF	113	105	92.92%	61.06%
	DF	152	87	57.23%	
	NF	17	16	94.11%	
	OC	67	0	0%	
IEC ratio	TF	113	106	93.805%	62.26%
	DF	152	84	55.26%	
	NF	17	17	100%	
	OC	67	0	0%	
Duval Triangle	TF	113	109	96.46%	70.03%
	DF	152	127	83.55%	
	NF	17	17	100%	
	OC	67	0	0%	
Key Gas	TF	113	62	54.86%	49.0%
	DF	152	98	64.67%	
	NF	17	13	76.47%	
	OC	67	0	0%	

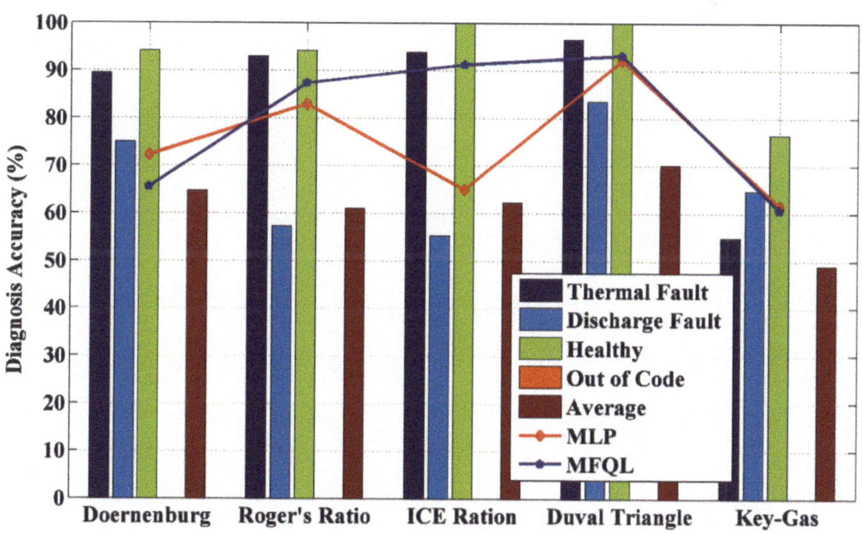

FIGURE 3.8 Conventional methods based accuracy analysis of DGA performance.

TABLE 3.6 MLP based performance analysis of DGA interpretation for transformer.

MLP model/ Data sets used	Type of transformer fault	Training phase performance		Diagnosis accuracy in training phase (%)		Diagnosis accuracy in testing phase (%)	
		MSE	RMSE				
MLP model 1: based on Doernenburg ratio [552×4]	HI	0.2750	0.5244	14/26	72.5	01/05	72.3
	PD			21/35		35/48	
	D1			47/52		26/33	
	D2			04/28		01/11	
	TF(<700°C)			53/75		44/55	
	TF(>700°C)			112/130		42/54	
MLP model 2: based on Roger's ratio [552×3]	HI	0.1680	0.4099	10/26	83.2	03/05	83.0
	PD			24/35		39/48	
	D1			49/52		28/33	
	D2			11/28		01/11	
	TF(<700°C)			68/75		50/55	
	TF(>700°C)			126/130		50/54	
MLP model 3: based on IEC ratio [552×3]	HI	0.286	0.5348	11/26	71.4	03/05	65.0
	PD			22/35		13/48	
	D1			44/52		16/33	
	D2			02/28		04/11	
	TF(<700°C)			63/75		44/55	
	TF(>700°C)			105/130		54/54	
MLP model 4: based on Duval Triangle [552×3]	HI	0.064	0.2523	24/26	93.6	03/05	92.2
	PD			33/35		47/48	
	D1			47/52		31/33	
	D2			19/28		08/11	
	TF(<700°C)			75/75		55/55	
	TF(>700°C)			126/130		46/54	
MLP model 5: based on Key Gas [552×7]	HI	0.292	0.5404	22/26	70.8	03/05	61.7
	PD			10/35		13/48	
	D1			20/52		12/33	
	D2			07/28		03/11	
	TF(<700°C)			56/75		42/55	
	TF(>700°C)			130/130		54/54	

4.2.2 *MFQL based power transformer fault diagnosis*

Based on conventional DGA interpretation methods, five different models have been implemented by using multi-class MFQL under as listed in Table 3.7. After comparing Table 3.7 with Table 3.5, it has been found that multi-class MFQL based DGA interpretation accuracy has been enhanced as compared to the conventional DGA interpretation accuracy. The highest diagnosis accuracy to be achieved by multi-class MFQL model#4 (Based on Duval triangle) is 97.69% (for training phase) and 93.20% (for testing phase). Diagnosis accuracy achieved is not desirable. Therefore a method is developed to solve enhance diagnosis accuracy.

TABLE 3.7 Multi-class MFQL based performance analysis of various DGA interpretation methods.

MFQL model/ Data sets used	Type of transformer fault	Training phase performance		Diagnosis accuracy in testing phase (%)		Diagnosis accuracy in testing phase (%)	
		MSE	RMSE				
MFQL model 1 based on Doernenburg ratio [552×4]	HI	0.1532	0.3914	23/26	84.68	03/05	65.53
	PD			19/28		02/11	
	D1			68/75		40/55	
	D2			110/130		34/54	
	TF(<700°C)			29/35		41/48	
	TF(>700°C)			44/52		15/33	
MFQL model 2 based on Roger's ratio [552×3]	HI	0.0954	0.3089	22/26	90.46	04/05	87.38
	PD			25/28		07/11	
	D1			66/75		50/55	
	D2			118/130		44/54	
	TF(<700°C)			35/35		46/48	
	TF(>700°C)			47/52		29/33	
MFQL model 3 based on IEC ratio [552×3]	HI	0.0723	0.2689	23/26	92.77	03/05	91.26
	PD			26/28		07/11	
	D1			70/75		48/55	
	D2			119/130		51/54	
	TF(<700°C)			34/35		48/48	
	TF(>700°C)			49/52		31/33	
MFQL model 4 based on Duval Triangle [552×3]	HI	0.0231	0.1520	25/26	97.69	04/05	93.20
	PD			26/28		10/11	
	D1			73/75		48/55	
	D2			129/130		53/54	
	TF(<700°C)			35/35		46/48	
	TF(>700°C)			50/52		31/33	
MFQL model 5 based on Key Gas [552×7]	HI	0.3844	0.62	17/26	61.56	03/05	60.68
	PD			20/28		07/11	
	D1			53/75		32/55	
	D2			70/130		34/54	
	TF(<700°C)			22/35		33/48	
	TF(>700°C)			31/52		16/33	

5 Implementation of AI methods based on proposed most relevant input variables

Here, in this section, again two AI approaches (MLP-ANN and MFQL) have been implemented based on selected most relevant input variables by the J48 and the PCA algorithm as well as based on all 24 variables as shown in input vector H and the obtained results are explained further.

5.1 MLP-ANN based proposed approach implementation

Based on the proposed approach, three models have been implemented and analyzed by using an MLP neural network as listed in Table 3.8. After analyzing the Table 3.8, it is concluded that MLP model#3 (based on selected variables by J48 algorithm) is the best-suited model for step-up transformer fault diagnosis, which has diagnosis accuracy of 97.4% (for training phase) and 97.1% (for testing phase). As compared with other MLP models (Table 3.6), the proposed approach gives suitable results for step-up transformer fault diagnosis. The graphical representation of MLP model#3 has been represented in Figs. 3.9–3.13 for further validation of the proposed approach performance.

TABLE 3.8 MLP based performance analysis for step-up transformer using proposed approach.

MLP model/ Data sets used	Type of transformer fault	Training phase performance		Diagnosis accuracy in training phase (%)		Diagnosis accuracy in testing phase (%)	
		MSE	RMSE				
MLP model 6 based on all variables [552×24]	HI	0.0950	0.3082	24/26	90.5	03/05	89.3
	PD			33/35		47/48	
	D1			48/52		31/33	
	D2			20/28		07/11	
	TF(<700°C)			75/75		51/55	
	TF(>700°C)			113/130		45/54	
MLP model 7 based on PCA [552×8]	HI	0.0550	0.2345	24/26	94.5	05/05	94.2
	PD			33/35		42/48	
	D1			48/52		30/33	
	D2			21/28		10/11	
	TF(<700°C)			75/75		55/55	
	TF(>700°C)			126/130		52/54	
MLP model 8 based on J48 [552×8]	HI	0.0260	0.1612	26/26	**97.4**	05/05	**97.1**
	PD			32/35		45/48	
	D1			50/52		31/33	
	D2			26/28		11/11	
	TF(<700°C)			75/75		55/55	
	TF(>700°C)			128/130		53/54	

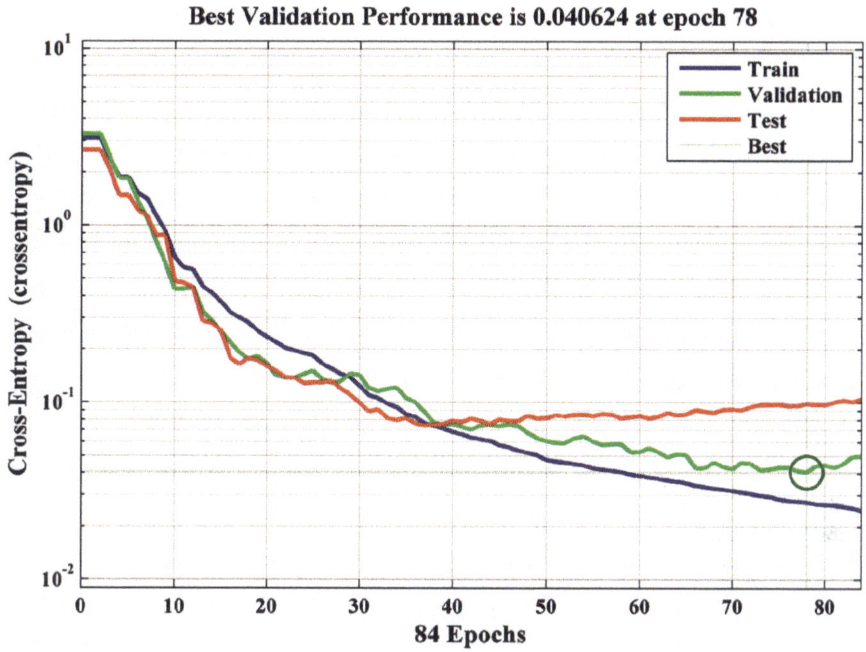

FIGURE 3.9 Performance plot of training phase.

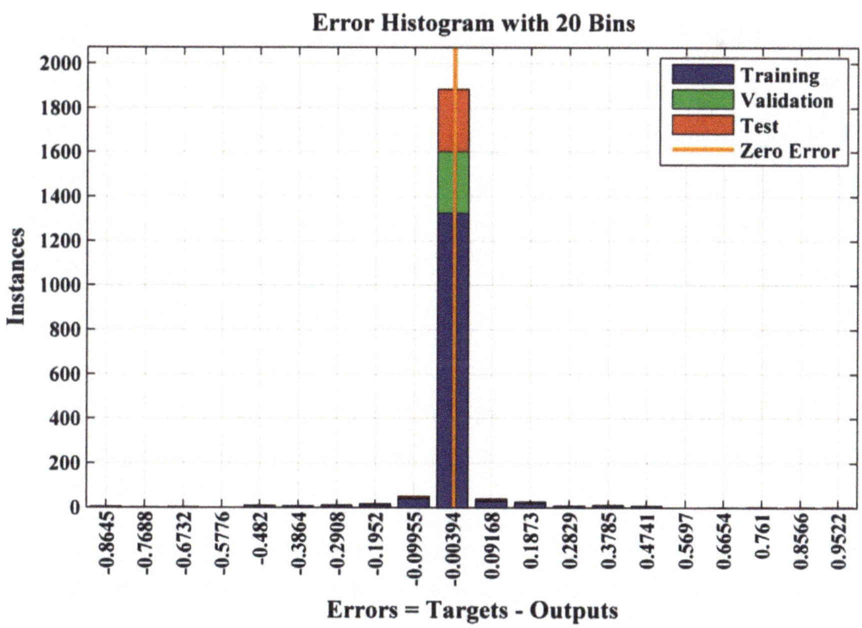

FIGURE 3.10 Error histogram plot of training phase.

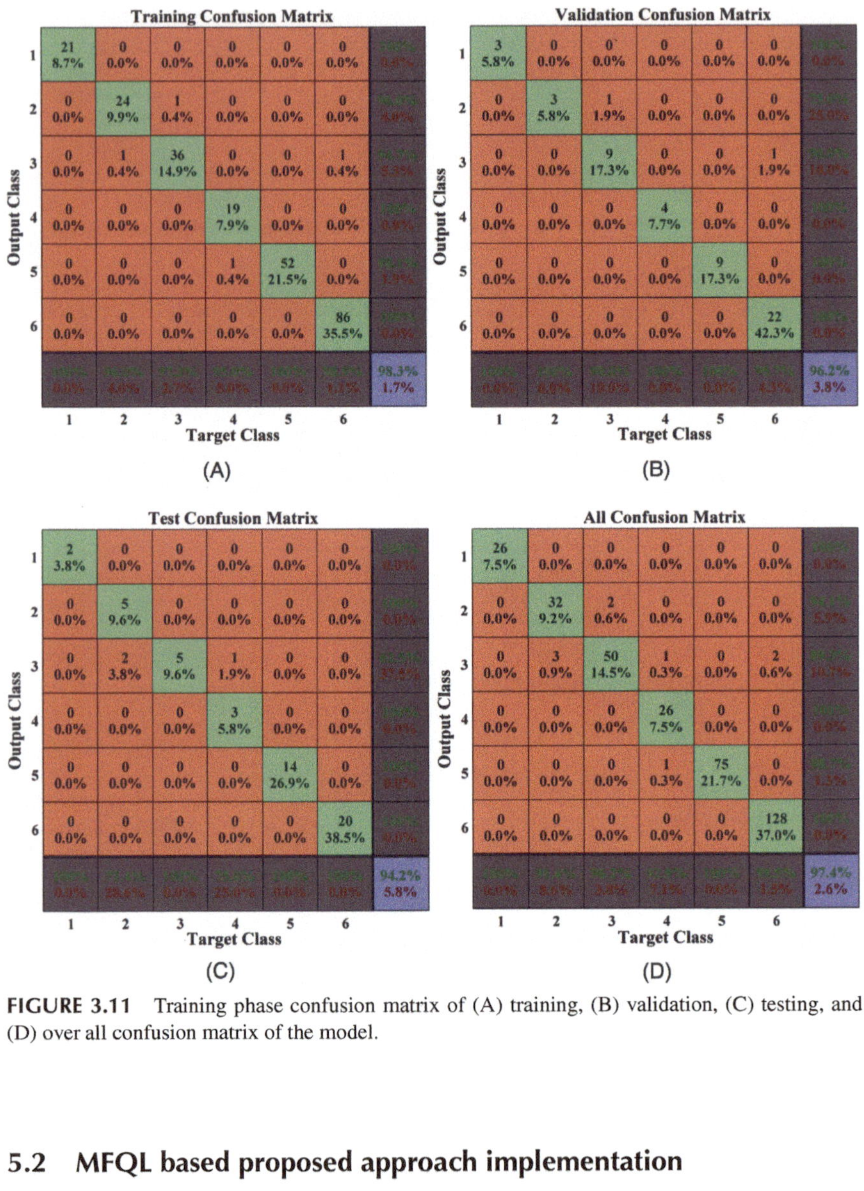

FIGURE 3.11 Training phase confusion matrix of (A) training, (B) validation, (C) testing, and (D) over all confusion matrix of the model.

5.2 MFQL based proposed approach implementation

Based on the proposed approach, three models have been implemented and analyzed by using multi-class MFQL as listed in Table 3.9. After analyzing the Table 3.9, it is concluded that MFQL model#3 (based on selected variables by J48 algorithm) is a good model for step-up transformer fault diagnosis, which has diagnosis accuracy of 99.71% (for training phase) and 98.06% (for testing phase). By comparing with other MFQL models (Table 3.7), the proposed approach gives more suitable results for step-up transformer fault diagnosis.

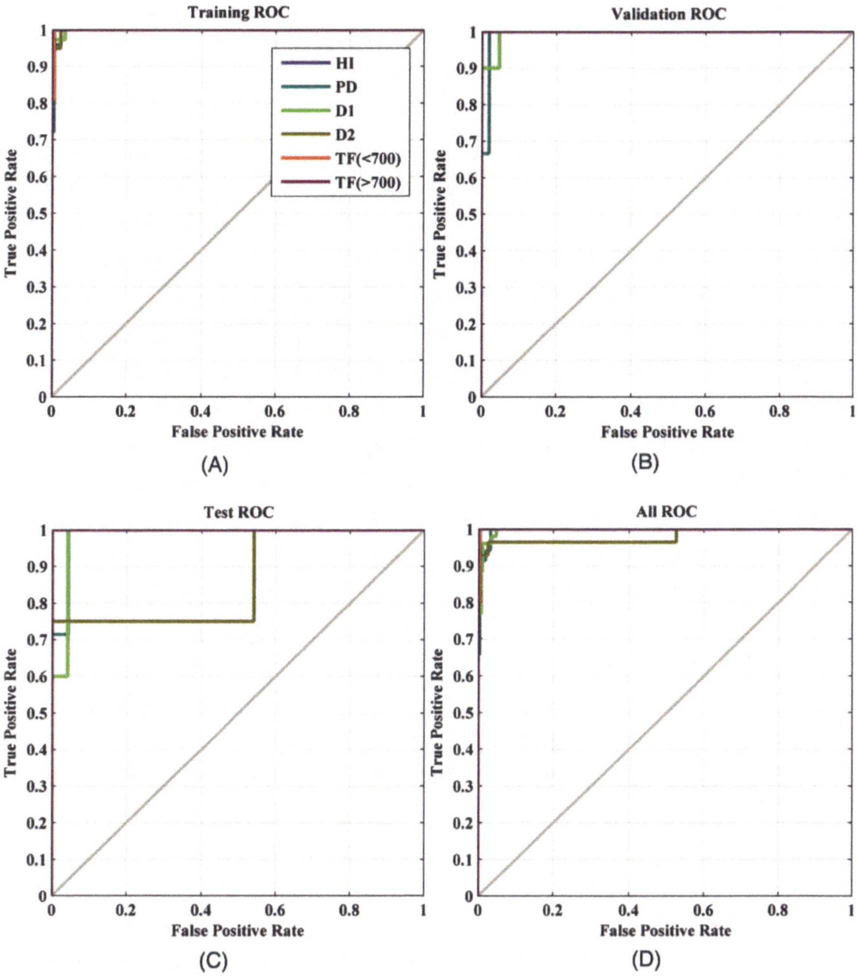

FIGURE 3.12 Training phase ROC plot of (A) training, (B) validation, (C) testing, and (D) over all ROC of the model.

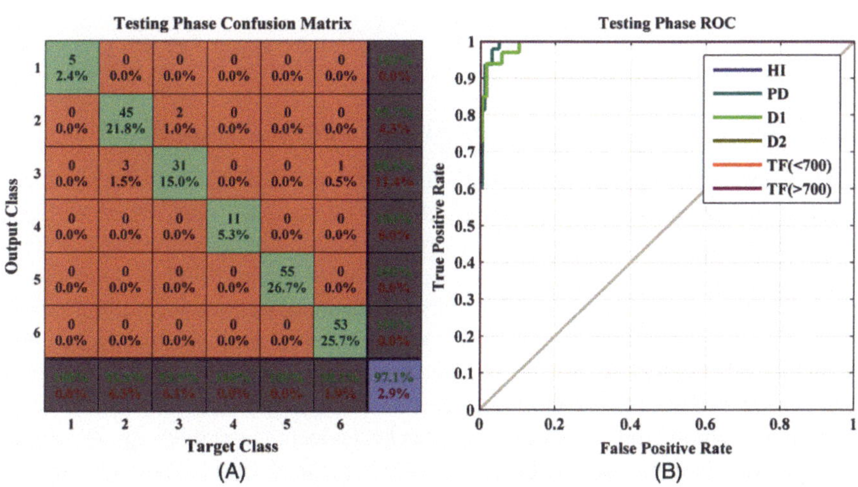

FIGURE 3.13 Testing phase plot of (A) testing phase confusion matrix and (B) ROC curve of the model.

TABLE 3.9 Multi-class MFQL based performance analysis for step-up transformer using proposed approach.

MFQL model/ Data sets used	Type of transformer operating condition	Training phase performance		CI/TS* Successful classification during training phase (%)		CI/TS* Successful classification during testing phase (%)	
		MSE	RMSE				
MFQL model 1: based on all variables [552×24]	HI	0.0318	0.1783	23/26	96.82	04/05	95.63
	PD			24/28		09/11	
	D1			74/75		52/55	
	D2			130/130		54/54	
	TF(<700°C)			35/35		47/48	
	TF(>700°C)			49/52		31/33	
MFQL model 2: based on PCA [552×8]	HI	0.0087	0.0933	26/26	99.13	04/05	97.09
	PD			27/28		09/11	
	D1			75/75		53/55	
	D2			130/130		54/54	
	TF(<700°C)			35/35		48/48	
	TF(>700°C)			50/52		32/33	
MFQL model 3: based on J48 [552×8]	HI	0.0029	0.0539	26/26	**99.71**	05/05	**98.06**
	PD			27/28		10/11	
	D1			75/75		55/55	
	D2			130/130		53/54	
	TF(<700°C)			35/35		48/48	
	TF(>700°C)			52/52		31/33	

5.3 Comparative analysis using AI based proposed approach

Two different AI approaches named MLP-ANN and MFQL have been implemented and analyzed for eight different data set used as input variables (i.e., Durnenburg ratios, Roger's ratios, IEC ratios, Key-gases, Duval triangle's variables, prepared vector H of 24 variables, selected variables by PCA, and selected variables by J48 algorithm). The highest diagnostic accuracy for the training and testing phase has been represented in Table 3.10 for all AI approaches under eight different categories. After a detailed analysis of Table 3.10, it is found that testing phase classification accuracy of MFQL approach has higher than MLP-ANN among the all under category#8 (selected variables by J48 algorithm) with 98.06%, which is desirable and higher than MLP diagnosis models.

TABLE 3.10 Summary of developed model's performance analysis of DGA interpretation.

Input variables used as per	MLP-ANN accuracy (%)		MFQL accuracy (%)	
	Train	Test	Train	Test
Durnenburg	72.50	72.30	84.68	65.53
Roger	83.20	83.00	90.46	87.38
IEC	71.40	65.00	92.77	91.26
Duval	93.60	92.2	97.69	93.20
Key gases	70.80	61.70	61.56	60.68
Full data (24 inputs)	90.50	89.30	96.82	95.63
Selected by PCA	94.50	94.20	99.13	97.09
Selected by J48	97.40	97.10	99.71	98.06

6 Conclusions

In this chapter, condition monitoring of a step-up transformer of WECS using DGA dataset has been presented. The detailed analysis of conventional DGA methods and its classification capability have been presented for six different operating conditions of step-up transformer. The incipient fault diagnosis of a step-up transformer has been presented with two different artificial intelligence approaches (i.e., MLP-ANN, and MFQL) by using DGA dataset under eight different categories (i.e., inputs are as Durnenburg ratios, Roger's ratios, IEC ratios, Key-gases, Duval triangle's variables, prepared vector H of 24 variables, selected variables by PCA, and selected variables by J48 algorithm). The comparative analyzed results show that MFQL gives better results as compared to MLP-ANN based other AI technique with input variables selected by the J48 algorithm. At the last, a comparative study of J48 algorithm-based results with PCA algorithm-based results have also been presented which shows the superiority of the J48 method for step-up transformer incipient fault diagnosis.

References

[1] J.A. Andrawus, Maintenance Optimization for Wind Turbines, A Ph.D thesis, The Robert Gordon University Aberdeen, 2008. doi: https://core.ac.uk/download/pdf/1576177.pdf.

[2] IEEE Guide for the Interpretation of Gases Generated in Oil-Immersed Transformers, ANSI/IEEE Std. C57.104–1991, 1992. doi: 10.1109/IEEESTD.1992.106973.

[3] IEC Guide for the Interpretation of Dissolved and Free Gases Analysis, IEC Std. IEC/CEI 60599, 2007.

[4] H. Malik, S. Mishra, Application of gene expression programming (GEP) in power transformer fault diagnosis using DGA, IEEE Trans. Ind. Appl. 52 (6) (2016) 4556–4565, doi: 10.1109/TIA.2016.2598677.

[5] X. Hao, S. Cai-xin, Artificial immune network classification algorithm for fault diagnosis of power transformer, IEEE Trans. Power Deliv. 22 (2) (2007) 930–935, doi: 10.1109/TPWRD.2007.893182.

[6] L.V. Ganyun, et al., Fault diagnosis of power transformer based on multi-layer SVM classifier, Int. J. Electr. Power Syst. Res. 74 (2005) 1–7, doi: 10.1016/j.epsr.2004.07.008.

[7] W.H. Tang, et al., A probabilistic classifier for transformer dissolved gas analysis with a particle swarm optimizer, IEEE Trans. Power Deliv. 23 (2) (2008) 751–759, doi: 10.1109/TPWRD.2008.915812.

[8] V. Duraisamy, et al., Neuro fuzzy schemes for fault detection in power transformer, Int. J. Appl. Soft Comput. 7 (2007) 534–539, doi: 10.1016/j.asoc.2006.10.001.

[9] R. Naresh, V. Sharma, M. Vashisth, An integrated neural fuzzy approach for fault diagnosis of transformers, IEEE Trans. Power Deliv. 23 (4) (2008) 2017–2024, doi: 10.1109/TPWRD.2008.2002652.

[10] A.R.G. Castro, V. Miranda, Knowledge discovery in neural networks with application to transformer failure diagnosis, IEEE Trans. Power Sys. 20 (2) (2005) 717–724, doi: 10.1109/TPWRS.2005.846074.

[11] H. Malik, S. Mishra, A.P. Mittal, Selection of most relevant input parameters using Waikato environment for knowledge analysis for gene expression programming based power transformer fault diagnosis, Int. J. Electric Power Components Syst. 42 (16) (2014) 1849–1862 http://dx.doi.org/10.1080/15325008.2014.956952.

[12] H. Malik, S. Mishra, Selection of most relevant input parameters using principle component analysis for extreme learning machine based power transformer fault diagnosis model, Int. J. Electric Power Components Syst. 45 (12) (2017) 1339–1352 https://doi.org/10.1080/15325008.2017.1338794.

[13] H. Malik, R. Sharma, Transmission line fault classification using modified fuzzy Q learning, IET Generat. Transm. Distrib. 11 (16) (2017) 4041–4050, doi: 10.1049/iet-gtd.2017.0331.

[14] H. Malik, S. Mishra, Artificial neural network and empirical mode decomposition based imbalance fault diagnosis of wind turbine using TurbSim, FAST and Simulink, IET Renew. Power Generat. 11 (6) (2017) 889–902, doi: 10.1049/iet-rpg.2015.0382.

[15] Parul Arora, et al., Wind Energy Forecasting Model for Northern-Western Region of India Using Decision Tree and MLP Neural Network Approach, Interdisciplinary Environmental Review 19 (1) (2018) 13–30, doi: 10.1504/IER.2018.089766.

[16] Nuzhat Fatema, et al., Data-Driven Occupancy Detection Hybrid Model Using Particle Swarm Optimization Based Artificial Neural Network, Metaheuristic and Evolutionary Computation: Algorithms and Applications, under Book Series "Studies in Computational Intelligence" (2020) 283–297, doi: 10.1007/978-981-15-7571-6_13.

[17] A Khatri, et al., Probabilistic Neural Network Based Incipient Fault Identification Using DGA Dataset, Procedia Computer Science 58 (2015) 665–672, doi: 10.1016/j.procs.2015.08.086.

[18] Savita Singh, et al., Application of Artificial Neural Network for Long Term Wind Speed Prediction, In Proc. IEEE CASP-2016 (2016) 217–222, doi: 10.1109/CASP.2016.7746168.

[19] Abdul Azeem, et al., Artificial Neural Network Based Intelligent Model for Wind Power Assessment in India, In Proc. IEEE PIICON-2016 (2016) 25–27, doi: 10.1109/POWERI.2016.8077305.

Chapter 4

Intelligent Data Analytics for 3-Phase Induction Motor Fault Diagnosis Using Gene Expression Programming (GEP)

1 Introduction

Over the last few decades, electrical motors have become an indispensable part of our lives; industries, transportation, aviation, agriculture, or even our day-to-day life. Even in this IT enabled modern era, motors have held on to their position as the prime movers of humanity. Among various types of electric motors, induction motors (IMs) are the most extensively used motor due to its various tremendous advantages. IMs have dominated all other motors because of some very desirable properties, as well as of its industrial, commercial as well as domestic level of applications. Out of several advantages, some of them are low operational cost, sturdiness, high efficacy and trustworthiness, low operation & maintenance (O&M) cost, excellent starting torque, and advanced level variable frequency drive (VFD) availability [1,2].

As a result, several electric motors used for the smart grid applications for commercial and domestic point of view are either 1-phase or 3-phase IMs [3]. As a result, to maintain the healthy operation for a particular application, the condition monitoring (CM) of IMs become more challenging and to reduce the downtime due to IMs malfunctioning, numerous CM approaches at advanced level are being proposed and implemented at real-site operation. Generally, faults/anomaly conditions in the IMs can be broadly categorized into three main groups (i.e., electrical, mechanical faults, environmental and maintenance fault) as shown in Fig. 4.1. Conventional CM techniques are based on microcontroller to detect the anomaly condition, whereas these approaches are not effective as much as it is required for practical application to reduce the maintenance time, enhance the reliability, and maintain the proper predictive maintenance (PdM) within a limited time period. So, to enhance the performance of all these application in the smart grid application, AI/machine learning approaches for intelligent data analytics based have been implemented [3].

Intelligent Data-Analytics for Condition Monitoring. http://dx.doi.org/10.1016/B978-0-323-85510-5.00004-1

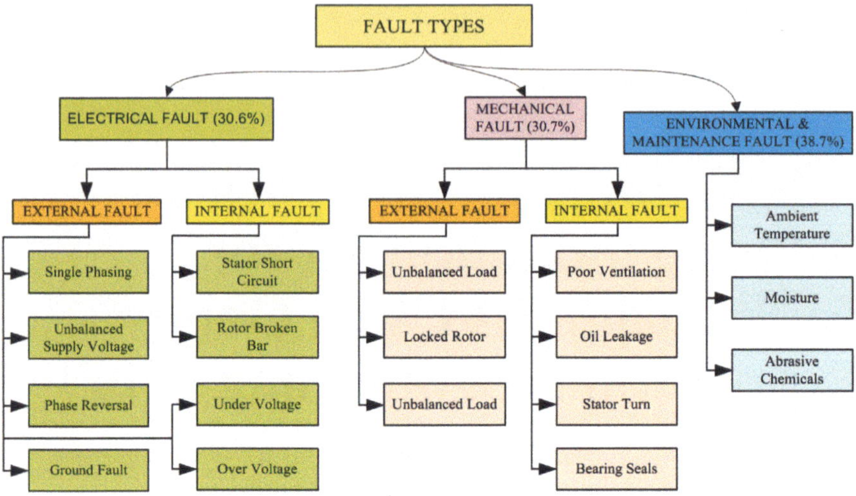

FIGURE 4.1 Fault classification tree for IMs.

Few of the anomaly condition or faults befalling in IMs are stator side faults [4–15], rotor side faults [16–27], bearing side faults [28–32], broken bars fault [33–36], static air gap eccentricity [37], stator winding faults [38]. Most of these faults are internal faults. But there are several external faults that affect the performance of IMs. Due to these external faults, smart grid application may get huge amount of loss. In this study, several external faults are analyzed and are listed in Table 4.1 for IMs failure cases.

Several external fault identification approaches for healthy and faulty condition evaluation, are performed based on mechanical as well as electrical signatures. Some of utilized signatures are as follow: vibration, electromagnetic field, induced voltage, surge test, motor circuit analysis, acoustic emission, air-gap torque, instantaneous angular speed, and motor current signature analysis (MCSA) [39]. However, most of these approaches are based on mechanical sensors and require expert skills or considerable experience and do not provide stable result for identical fault condition. Accurate fault analysis is still a disputed domain in the area of FDCM of IMs. For appropriate classification of faults and for scaling earlier cited difficulties, many researchers have advocated use of different AI techniques, for example

TABLE 4.1 Codification of the faults experienced by 3-ph IMs.

Fault types	Healthy condition	Single phase	Unbalanced-voltage	Under-voltage	Over-voltage	Locked rotor	Overload
Code	NF	*SP*	*UB*	*UV*	*OV*	*LR*	*OL*

Refs. [1–3,40–44] and developed precise diagnostic procedures based on current signature analysis.

In Refs. [1,2,40], artificial neural network (ANN) applied with back propagation (BP) algorithm was used to diagnose IM faults. 3-phase RMS values of current and voltage as in Refs. [1,2] and negative sequence voltage as in Ref. [40] are used to identify the healthy condition of IM using BP algorithm on the ANN model. In Ref. [3], support vector machine (SVM) architecture was used for classifying external faults using RMS value of 3-phase voltage and current signature. Rodrigo H. Cunha Palácios and colleagues [41] present a comparative study of Naïve Bayes (NB), Decision Tree (C4.5), SVM, k-Nearest Neighbor (KNN), Repeated Incremental Pruning to Produce Error Reduction (RIPPER), and multilayer perceptron (MLP) for fault diagnosis of three different 1 hp IM.

ANN based approaches involve determination of local minima and an optimal network structure, both of which are time consuming processes with an added threat of over fitting [45]. SVM undergoes in complexity while selecting optimum parameters (*C* and lambda) and is prone to over fitting. Besides, the parameters of the target problems are also tough for interpretation [46].

In Ref. [42], external faults have been diagnosed using extreme learning machine (ELM) method, which is new and has faster processing. RMS values of voltages and current are taken as inputs to the ELM model. Main disadvantage of this approach is that it cannot encode more than one layer of abstraction and has very slow evaluation. For IM fault diagnosis problem, evaluation speed is rather more important that training and testing speeds.

In Ref. [43], the authors develop a multi-class fault diagnosis scheme for IM using Hilbert transform (HT) and Wavelet transform (WT). Vibration signals of sensor are used in the study to classify seven conditions of IM. Some drawbacks of WT are: WT decomposes signals into frequency bands and for controlling number of frequency bands; decomposition level setting needs to be set by the user in every run. In WT, the analysis is dependent on the selection of the wavelet basis function which is achieved by years of experience. However, WT method fails to split the high frequency bands where modulation information of the machine faults exists. In Ref. [44], author proposes a novel approach based on proximal support vector machine (PSVM) for diagnosing external faults. However, selection of optimal parameters in PSVM is a challenging task. Hence, we need optimization techniques (i.e., genetic algorithm-GA, genetic programming-GP, etc.) for optimal parameter selection leading to higher computational burden.

The GEP approach for fault identification is able to overcome all of these drawbacks. In general, GEP is able to garner best generalization than the MLP, SVM, PSVM, WT, ELM, GA, and GP. The main objective of this work is to propose a GEP based technique which can achieve fault classification accuracy as GEP combines the advantage of both GA and GP algorithms [47]. Our proposed approach does not require any additional optimization techniques, for example, ANN, SVM, WT, PSVM, and ELM etc. Ours is a first attempt at applying GEP for fault diagnosis and condition monitoring of 3-phase IMs.

This paper has five sections as follows: in Section 1, introduction along with literature review is presented. Datasets used in the research and methodology applied are explained in Section 2. Proposed fault diagnosis and condition assessment (FDCA) model formulation is presented in Section 3. Section 4 presented the results and conclusions are described in Section 5.

2 Brief information for IM condition monitoring innovations

The area of condition monitoring of the IM is a very mature area throughout the globe. Numerous researchers and academician are working in this field as per the Industry 4.0 demands. One data analytics study has been performed for the research area of the CM of IM by utilizing "The Lens" platform [48]. The Lens platform is an open access platform to provide the facility for patent search, which is launched in 2000 and it is allowed for free search over the 10 million patent documents. Based on the search, following information have been achieved in the domain of IM CM and FDD.

As per the available data, two type of analysis are performed: (1) Scholarly work based analysis and (2) Citing patents. So, in the innovation analysis, following are the top organizations: Ford Global Technologies LLC, General Electric Company, Samsung Electronics Co. Ltd, Ethicon LLC, Applied Materials Inc, Halliburton Energy Services Inc, At&t Intellectual Property I.L.P., GM Global Technology Operations LLC, Ford Motor Company, and The Regents Of The University Of California. The top inventors of this area are: Silverbrook Kia, Hyde Roderick A, Henry Paul Shala, Shelton Iv Frederick E, Gerszberg Irwin, Barzegar Farhad, Willis III Thomas M, Wood Jr Lowell L, Bennett Robert, Barnickel Donald J, Lapstun Paul, Ruben Steven M, Rosen Craig A, Bevec Dorian, Tang Y Tom, Kare Jordin T, Golz Stefan, Medoff Marshall, Harris Jason L, and Cavalli Fabio. According to the rank in the region as per the number of jurisdiction are: United States, WO—WIPO, Australia, European Patents, China, Japan, Germany, United Kingdom, France, Korea Republic of Canada, Russia, Finland, Brazil, New Zealand, Singapore, Hong Kong, Taiwan, Latvia, Eurasian Patents, Italy, Mexico, Malaysia, Spain, Argentina, and Romania. According to the top CPC classifications of the innovation area, the following research area are as per the rank: A61K38/00, C07K14/47, C07K14/705, C12Q2600/158, C12Q1/6883, A61K48/00, C07K2319/00, A61K2300/00, A61K45/06, A01K2217/05, G01N33/6896, G01N2500/00, A61K2039/505, G01R31/343, G01N33/5008, A61N1/36007, C07K16/18, C12Q2600/136, A01K2227/105, G01N33/6893. For the full explanation of all CPC classifications, reader may refer The Lens.

Moreover, the condition monitoring market in the globe is very big of around 12.8 billion and 13.08 billion in 2017 and 2018, respectively, and it is forecasted around 14.6–15.1 billion by 2026. According to the market breakdown, North America is at the top followed by Latin America, Europe, Middle East & Africa, and Asia Pacific region. The growth rate of the IM in 2020 is very low due to pandemic situation but it is expected to grow very big in future. Moreover, in India the electric motor market is also very big in term of industrial uses. Therefore it is necessary to have a good CM and FDD for the healthy operation of IM.

3 GEP methodology and data sources

3.1 Database used for study

A 788 dataset, which is publicly available in Ref. [3], has been utilized for developing GEP based external fault identification approach for 3-phase IMs. The dataset comprises of data samples for seven distinct conditions (i.e., six external fault conditions and a normal operating scenario). After successful training and testing using this dataset, developed model is used to diagnose external faults simulated on 3-phase IM in lab. The experimental setup used to validate the identifier, has been developed in Lab of NIT Hamirpur, H.P. India during a post graduate research project as shown in Fig. 4.2. External faults were simulated on a 1/3 hp 3-phase squirrel-cage IM (SCIM). In recorded datasets, simulated external faults were same as publicly available dataset of Ref. [3], that is, SP, UB, UV, OV, LR, OL. The UB, UV, and OV faults were simulated by using three different variable Variacs. The OL and LR faults were simulated by using a pony brake system and the SP fault was simulated by disconnecting 1-phase of the power line. Voltage and current data was collected for these simulations. Sampling frequency of 1.0 kHz was used to obtain 60 k samples data points for seven cases:

FIGURE 4.2 **Test setup for data aquisition in the laboratory.**

one healthy and six faulty conditions. For further validation, 900 sample data points are used in this study, out of them some records are presented in Table 4.5.

3.2 Gene expression programming (GEP)

The GEP is an advance data analytics approach. Basically, GEP is the combination of the advantages of GA and GP approaches. So, GEP has both advantages of GA and GP [47]. It is invented in 1999 by Candida Ferreira [49] for the several applications. In the modern Industry 4.0 domain, it is being utilized widely for the classification, forecasting, prediction, pattern recognition, optimization, feature selection, etc. The basic information for the GEP implementation is represented in Fig. 4.3, which includes all three procedures (i.e., GEP genes, the formation of the expression tree from the genes, and finally, representation of mathematical expression for easy of understanding). The mathematical implementation of fitness function used in GEP (f_i) for an individual program (i) is represented in Eqs. (4.1) and (4.2) [49]. The complete working architecture for GEP implementation is represented in Fig. 4.4.

$$f_i = \sum_{q=1}^{\lambda_t} \left(R - \left| \lambda_{(p,q)} - A_q \right| \right) \tag{4.1}$$

$$f_i = \sum_{q=1}^{\lambda_t} \left(R - \left| \frac{\lambda_{(p,q)} - A_q}{A_q} \times 100 \right| \right) \tag{4.2}$$

Where R = selection range, $\lambda_{p,q}$ = predicted value of the individual program (p) for fitness case (q) out of n fitness cases, and A_q = target value for fitness case q. For ideal fit, $\lambda_{p,q} = A_q$ for all fitness cases and $f_i = f_{\max} = n.R$.

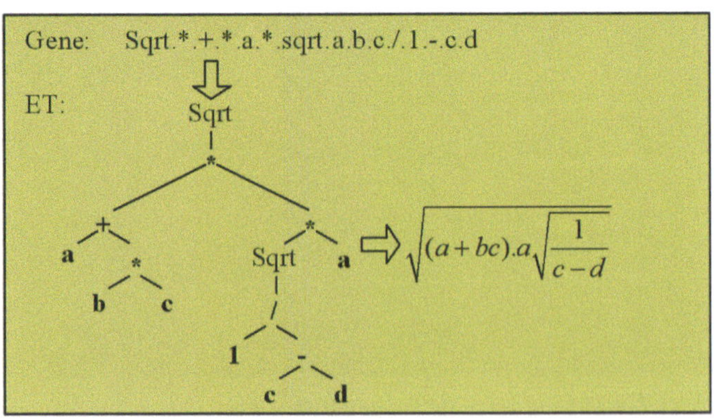

FIGURE 4.3 Graphical example for GEP and it's ET associated with mathematical expression.

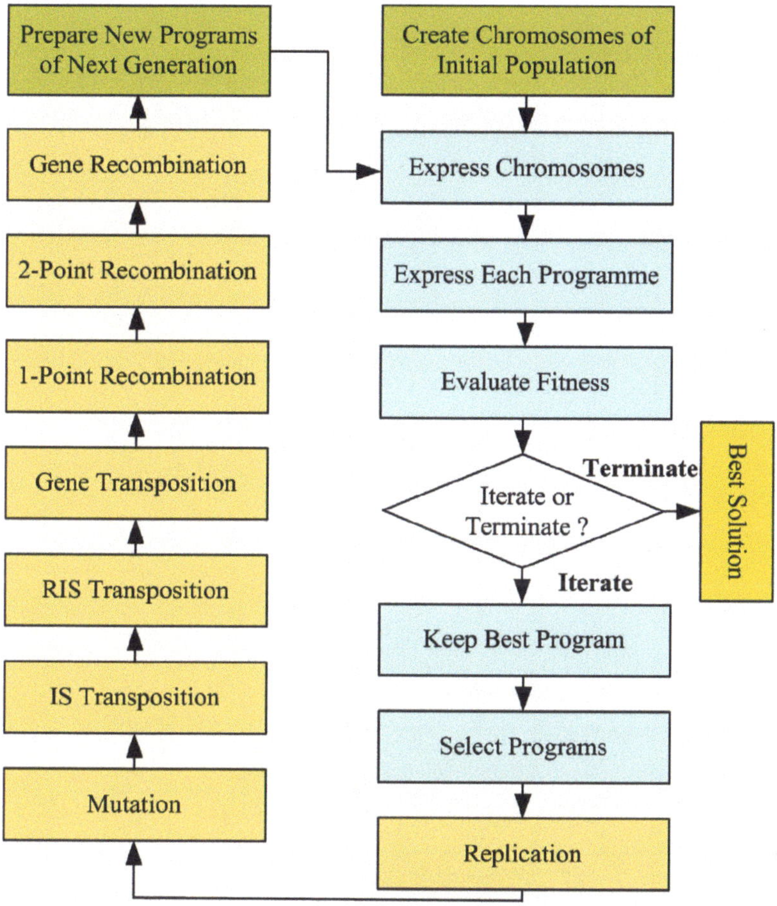

FIGURE 4.4 **Working steps for GEP approach [49].**

4 External fault classifier based on GEP

4.1 Dataset: training and testing

Most relevant input variables in the dataset have been collected as explained in section 3.1, we divided data into 2 sub-datasets: randomly chosen from 628 cases as training data set and the remaining 160 as the testing data set. All 788 cases have been analyzed and resulting identification of 7-classes have been presented in Table 4.2. The GEP FDD (fault detection and diagnosis) model is implemented based on available data sets which serve as input signature during training and testing phases.

4.2 The GEP approach

Our proposed approach has two basic steps as shown in Fig. 4.5: reading the input parameters and implement the improvements. The GEP uses simulated experimental results in lab and publicly available online datasets of Ref. [3].

TABLE 4.2 Dataset used for model formulation.

Dataset	(NF)	(*SP*)	(*UB*)	(*UV*)	(*OV*)	(*LR*)	(*OL*)	Grand total
Training	123	68	358	39	8	8	24	628
Testing	31	17	92	10	2	2	6	160
Sub-total	154	85	450	49	10	10	30	788
Simulated sample in lab for study	190	116	441	60	27	22	44	900

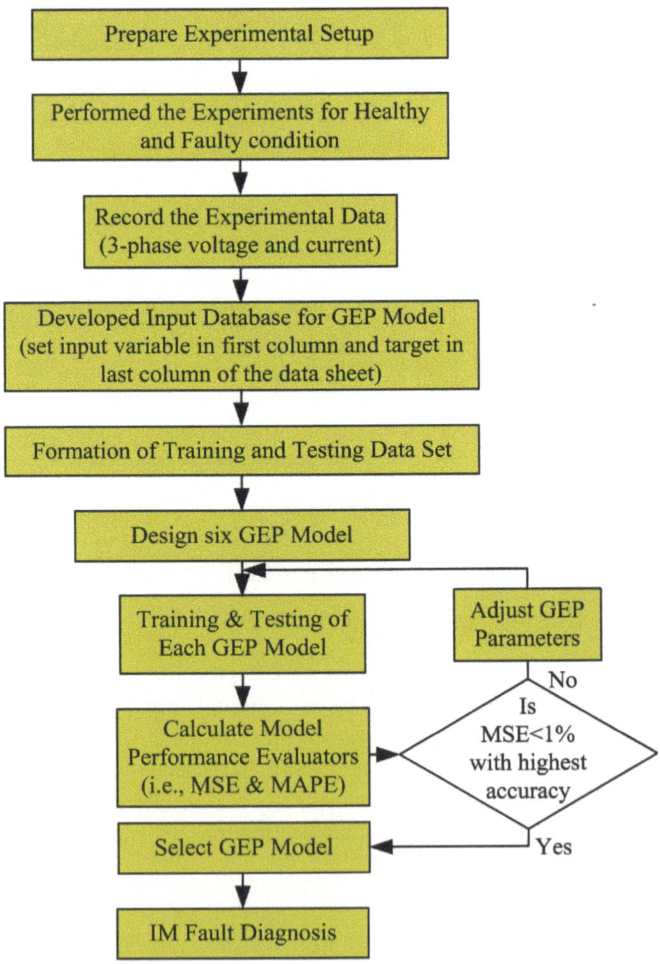

FIGURE 4.5 Proposed approach implementation for IM fault diagnosis.

For the first step, datasets have been collected from two sources: experimental datasets which are simulated in lab and publicly available dataset. The model is designed on the publicly available dataset and then tested with experimental datasets. In the further step, six binary GEP models are formulated to classify six external faults along with healthy scenario. Each GEP model is being trained and later performance measures (i.e., MSE, RMSE, and MAPE) are calculated to analysis the external fault classification accuracy of the model.

4.3 GEP model formulation

The GEP model is formulated for the demonstration point of view based on Fig. 4.5, which is represented in Fig. 4.6. The developed model includes six-binary GEP classifiers, which are being utilized to classify seven-conditions: one healthy condition (NF) along with six faulty conditions.

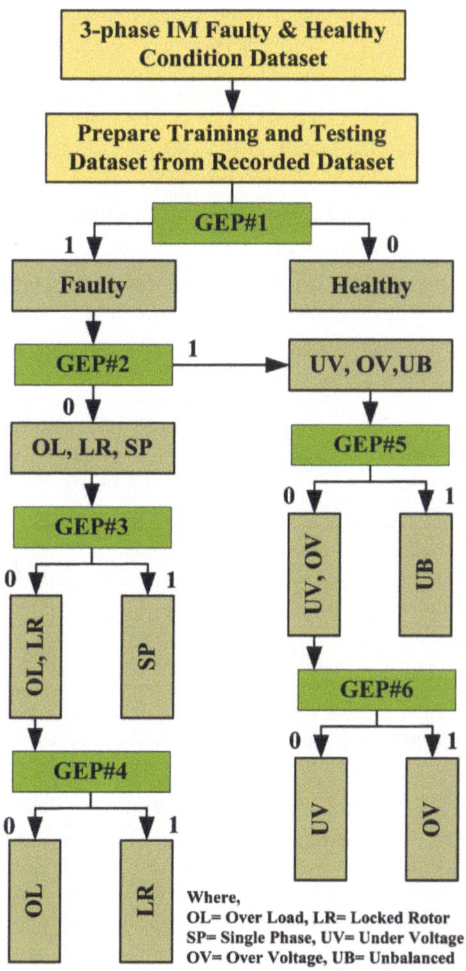

FIGURE 4.6 GEP based binary classifier demonstration for IM FDD.

For the demonstration of the GEP FDD performance, GEP#1 (model-1) is trained by utilizing the whole training dataset and segregate the healthy condition from the all faulty conditions. For this operation, the used code for healthy condition representation is 0. Now, GEP#2 (model-2) is trained by utilizing segregated faulty data from model-1 for six faults condition. The used code at the output side of GEP#2 is 0 for segregating the faults conditions "SP, LR, & OL" from the "UB, UV, & OV." Now, GEP#3 (model-3) is trained by utilizing the segregated faulty data of "SP, LR, & OL" obtained from model-2 for three faults condition. The used code at the output side of GEP#3 is 1 to segregate the faults conditions "SP" from the "LR & OL." Now, GEP#4 (model-4) is trained by utilizing the segregated faulty data of "LR & OL" obtained from model-3 for two faults condition. The used code at the output side of GEP#4 is 1 to segregate the faults conditions "LR" from the "OL." Now, GEP#5 (model-5) is trained by utilizing the segregated faulty data of "UB, UV, & OV" obtained from model-2 for three faults condition. The used code at the output side of GEP#5 is 1 to segregate the faults conditions "UB" from the "UV & OV." Finally, GEP#6 (model-6) is trained by utilizing the segregated faulty data of "UV & OV" obtained from model-5 for two faults condition. The used code at the output side of GEP#6 is 1 to segregate the faults conditions "OV" from the "UV." As a summary of all output codes used for representation of faulty condition at the output side of the GEP models are shown in Table 4.3:

We have used following optimized parameters for six GEP models:

For Fitness function:	Rounding threshold = 0.5, Inner margin = 1, Fitness function = Dual Margin, Outer margin = 2.5
For Numerical Constants:	Data Type = Floating-Point, Constants per Gene = 10, RNC Mutation = 0.00206, Upper Bound range = 10, Lower Bound range = −10, Dc Inversion =0.00546, Dc Mutation = 0.00206, Dc IS Transposition = 0.00546
For General setting:	Tail size = 9, Head size = 8, Chromosomes number = 30, Dc size = 9, Genes number = 3, Gene size = 26, Generation number = 1000 (maximum), Linking function = Addition
For genetic operator:	Inversion = 0.00546, Mutation = 0.00138, Transposition of RIS = 0.00546, IS Transposition = 0.00546, Recombination of One-point = 0.003, Recombination of Two-point = 0.003, Transposition of Gene = 0.00277, Recombination of Gene = 0.00277

The model's FDD performance of each GEP-classifier model (GEP 1–GEP 6) is analyzed by computing performance measures of the model after modification of necessary parameters.

TABLE 4.3 Used code for GEP diagnostic model demonstration.

Fault	NF	OL	LR	SP	UV	OV	UB
GEP#1	0	1	1	1	1	1	1
GEP#2		0	0	0	1	1	1
GEP#3		0	0	1			
GEP#4		0	1				
GEP#5					0	0	1
GEP#6					0	1	

Classification accuracy (CA) is computed as

$$CA = \frac{No.of\ correctly\ classified\ samples}{No.of\ total\ samples\ in\ the\ Dataset} \tag{4.3}$$

MSE is calculated as:

$$MSE = \frac{1}{n}\sum_{k=1}^{n}(E_k)^2; E_k = |D_k - T_k| \tag{4.4}$$

Where, n = number of data sample, D_k = target value, and T_k = model output. Mean absolute percentage error (MAPE),

$$MAPE = \frac{1}{n}\sum_{k=1}^{n}\left|\frac{Fault_{GEP\_classifier} - Fault_{actual}}{Fault_{actual}}\right| \times 100 \tag{4.5}$$

For the demonstration of the proposed approach performance evaluation, voltages and current signature are used [1–3]. Proposed GEP models have been developed by utilizing the GeneXproTools platform which is developed by Gepsoft Ltd., Bristol, United Kingdom [50] and its working procedure is shown in (Fig. 4.7). The models performance parameters are computed by utilizing the Eqs. (4.3)–(4.5), and evaluated results are tabulated in Table 4.4. Moreover, for the further one-step-ahead validation point of view, 17 new data samples have been analyzed and demonstrated results are represented in Table 4.5.

We calculate classification accuracy, MSE and MAPE corresponding to each GEP model as per Eqs. (4.3), (4.4) and (4.5) and are shown in Table 4.4 (Fig. 4.7).

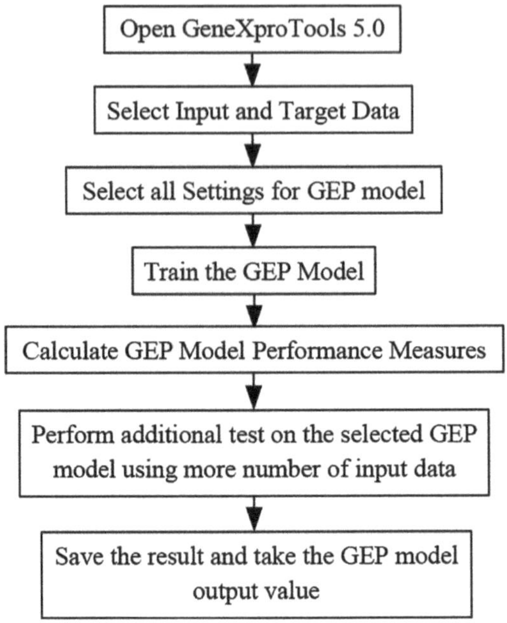

FIGURE 4.7 High level implementation procedure for GEP model.

TABLE 4.4 GEP model performance analysis for IM fault demonstration.

Developed models	Demonstrated performance representation				
	MAPE	MSE	RMSE	Accuracy (%)	CPU time (s)
GEP#1	0.18133	0.00126	0.03562	99.87	21.0
GEP#2	0.19591	0.003154	0.05617	99.68	13.52
GEP#3	0	0	0	100	8.3
GEP#4	0	0	0	100	4.0
GEP#5	0	0	0	100	3.80
GEP#6	0	0	0	100	3.0
Mean performance	0.581819	0.000736	0.01529	99.925	8.9367

5 Results and discussion

Fig. 4.5 shows the methodology implemented for classifications of external faults experienced by 3-phase IM using the GEP approach. Classification accuracy calculation is presented in this section. The overall efficiency of fault

TABLE 4.5 Historical data-based demonstration of the proposed approach.

S. no.	Input parameters						AFC*	GEP
	I_1	I_2	I_3	V_1	V_2	V_3		
1	0.4332	0.4234	0.4243	2.6992	2.6523	2.6981	NF	NF
2	0.4062	0.4310	0.4439	2.6980	2.6542	2.6832	NF	NF
3	0.6324	0.6358	0.0212	2.6768	2.6152	2.7108	SP	SP
4	0.6981	0.6932	0.00788	2.6862	2.6008	2.7912	SP	SP
5	0.252100	0.6620	0.77280	0.93346	2.6315	2.63401	UB	UB
6	0.545900	0.2583	0.64263	2.6316	1.9274	2.81201	UB	UB
7	0.324402	0.5454	0.27419	1.9262	2.61407	1.99116	UB	UB
8	0.243708	0.67458	0.451243	1.9634	2.8858	2.63774	UB	UB
9	0.942402	0.80744	0.181818	2.9031	2.8729	1.36432	UB	UB
10	0.245160	0.25012	0.233882	1.45368	1.54140	1.44311	UV	UV
11	0.296016	0.29616	0.29096	1.8845	1.8556	1.87413	UV	UV
12	0.46412	0.48670	0.46528	2.8833	2.8612	2.8947	OV	OV
13	1.42760	1.3866	1.43136	2.7733	2.62133	2.6957	LR	LR
14	0.80414	0.79340	0.78752	2.63866	2.60156	2.67461	OL	OL
15	0.854654	0.820242	0.844341	2.64841	2.60445	2.6844	OL	OL
16	0.28439	0.80401	0.20087	1.3971	2.01325	2.00372	UB	UB
17	0.419135	0.411307	0.432708	2.65540	2.46344	2.07118	NF	NF

classification using GEP was 99.925% (Table 4.5). Individual results for each GEP model are followed as:

GEP model-1: Classification accuracy of **99.87%:**

y = gepGOE2A((1.0-((d(V3)-d(V3))+(d(I2)-d(I1)))),(gepLT2 G(gepGOE2 B(d(I1),d(V2)),d(I1))^4));

y = y + gepGOE2E((gepGOE2G((d(I3)-d(I1)),gepGOE2A(d(V2),d(V3)))^4),g epGOE2G((G2C3^4),exp(G2C0)));

y = y + (gep3Rt(gepLT2C(gepLogi(gepGOE2A(G3C4,d(I2))), (G3C0+gepLT 2A(G3C1,d(V1)))))^3);

where, calculated value of G2C3 = -8.3969384746849;
G2C0 = -2.41074861903745; G3C0 = -2.18054750206;
G3C4 = -2.84707174901578; G3C1 = 9.53649006622516;

Translation of a GEP1 gene into an expression tree (ET1) is shown in Fig. 4.8:

GOE2A.NOT.X4. + .LT2G.-.-.GOE2B.d3.d2.d2.d4.d3.d3.d1 .d1.d4 +

GOE2E.X4.GOE2G.GOE2G.X4.Exp.-.GOE2A.c3.c0.d5.d3 .d1.d2.d4.c3.c0 +

X3.3Rt.LT2C.Logi. + .GOE2A.c0.LT2A.c4.d4.c1.d0.c5.d0.d1.d2.c9

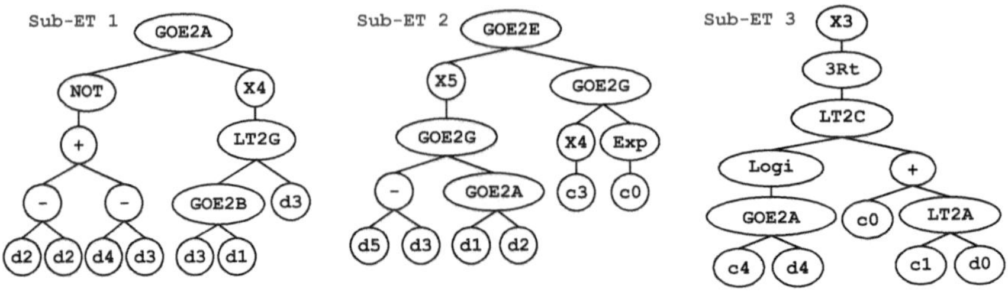

FIGURE 4.8 ET representation for GEP model#1.

GEP model-2: Classification accuracy of **99.68%**

y = (gepLT2A(atan(gepAND2((d(V1)-d(V3)),d(V1))),gepLT2 A(gepGOE2B (d(I2),d(I2)),(G1C1^3)))^2);

y = y + (atan(((d(V3)/d(I1))/G2C7))*gepGOE2G(gepGOE2E(d(I1),d(V3)), gepGOE2G(G2C7,G2C8)));

y = y + gepAND2((gepAND2(d(I3),d(I1))+gepAND2(G3C2, G3C2)),(d(V1)-(d(V2)^3)));

where, calculated value of G1C1 = 3.7137974181341; G2C7 = 5.79088717307047; G2C8 = 4.64339121677297; G3C2 = -5.62547685171056;

Translation of GEP2 gene into an expression tree (ET2) is given in Fig. 4.9:

X4.AND1.-.d2.*.AND2.d0.X3.d0.d3.d2.d0.d3.c9.d3.d1.d1 +

.AND2.GOE2C..OR2.LT2G.*.AND2.d0.d3.d4.d2.d1.d3.d1.d4.c4 +

X4.Avg2.X4.Sqrt.c2.OR2.*.Avg2.d5.d4.d0.d0.d2.d2.d5.d3.d0

GEP model-3: Classification accuracy of **100%**

y = gepAND1((gepLT2B((1.0-d(I2)),G1C4)/(d(V2)*d(I3))),(gepAND1(d(V3), G1C0) + (d(I1) + d(I2))));

y = y + G2C5;

y = y + gepLT2B((gepLT2G(gepLT2B(G3C0,G3C0),d(I1)) + (G3C4-G3C2)),gepGOE2G(gepLT2C(d (I1),d(I2)),(d(V3)*G3C 3)));

where, calculated value of G1C4 = -3.44401379436628; G1C0 = 9.68077639088107; G2C5 = -8.7204718161565; G3C4 = 4.51228064821314; G3C2 = 4.91500595110935; G3C3 = 6.22852259895627; G3C0 = 2.49305703909421;

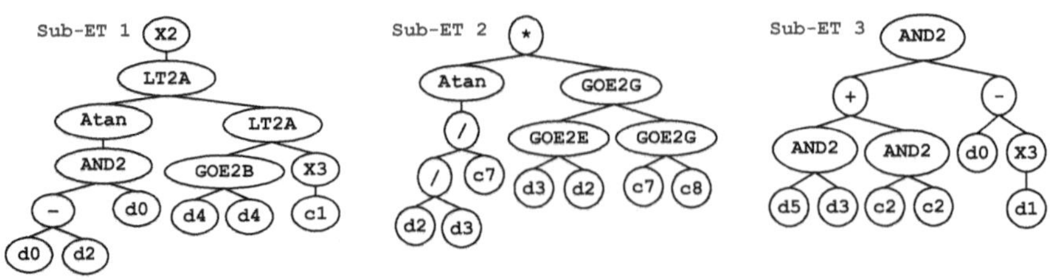

FIGURE 4.9 ET representation for GEP model#2.

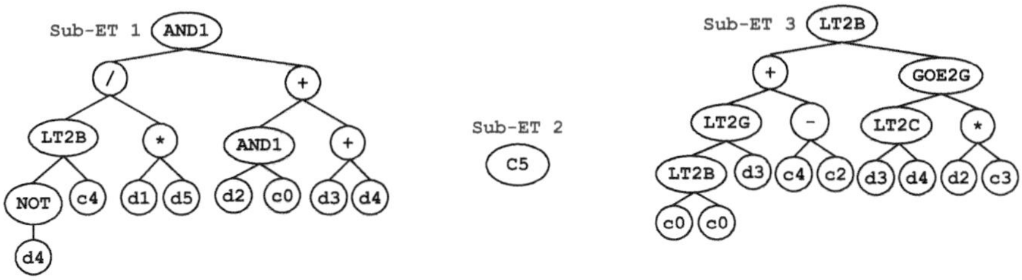

FIGURE 4.10 ET representation for GEP model#3.

Translation of a GEP3 gene into an expression tree (ET3) is shown in Fig. 4.10:

AND1./.+.LT2B.*.AND1.+.NOT.c4.d1.d5.d2.c0.d3.d4.d4.c1 +

c5.LT2A.NOT.c4.c7.d4.*.c4.d5.c4.c6.d4.d5.c0.c5.d5.d5 +

LT2B.+.GOE2G.LT2G.-.LT2C.*.LT2B.d3.c4.c2.d3.d4.d2 .c3.c0.c0

GEP model-4: Classification accuracy is **100%**

y = gepOR2((G1C5+gepGOE2A(atan(G1C8),d(I2))),gepLT2 B(G1C6,(d(I3)+G1C6)));

y = y + gepAND2(gepLT2A(((d(I3)+d(V3))^2),gep3Rt(d(I1))),gepLT2C(((G2C8+d(I1))/2.0),d(V3)));

y = y + gepOR1((gepLT2G((d(V2)-d(I1)),G3C7)+gepGOE2A (G3C9,G3C1)), gepOR1((d(V3)+d(V3)),(d (V1)+d(V2))));

where, calculated value of G1C5 = -2.37098300119022; G1C6 = 4.89791558580279; G1C8 = 7.90765099032563; G2C8 = -6.40308847315897; G3C7 = -0.262764366588336; G3C9 = -3.93963438825648; G3C1 = -4.9534592730491;

Translation of GEP4 gene into an expression tree (ET4) given in Fig. 4.11:

OR2.+.LT2B.c5.GOE2A.c6.+.Atan.d4.d5.c6.c8.c9.d3.c4.c9.d1

+

AND2.LT2A.LT2C.X2.3Rt.Avg2.d2.+.d3.c8.d3.d5.d2.d2.c0.d3.c6

+

OR1.+.OR1.LT2G.GOE2A.+.+.-.c7.c9.c1.d2.d2.d0.d1.d1.d3

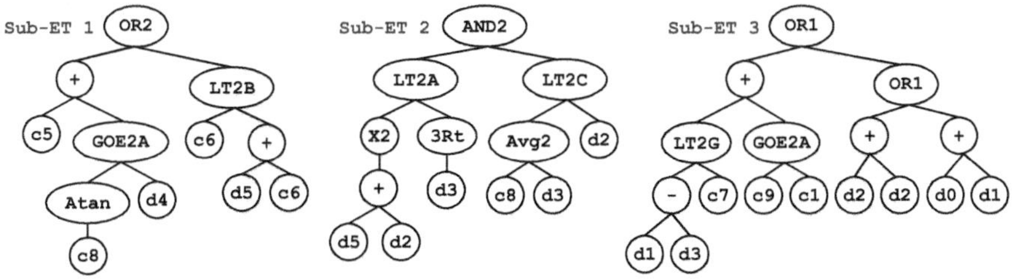

FIGURE 4.11 ET representation for GEP model#4.

GEP model-5: Classification accuracy is **100%**

y = gepLT2C(gepLT2E(gepGOE2B((G1C7^4),G1C7),(G1C5 *G1C4)),gepLT2E(gepOR2(d(3),G1C4), atan(G1C4)));

y = y + (gepGOE2B(((d(1)+G2C7)*(G2C7^2)),gepGOE2G(d(5),d(2)))^2);

y = y + gepGOE2B(((gepLT2C(d(5),d(6))*(d(3)^4))^4),gepO R2(gepGOE2B(d(2),G3C5),(d(6)^2)));

where, calculated value of G1C7 = -6.51173436689352; G1C5 = -5.75975829340495; G1C4 = -4.58440350910062; G2C7 = -52.5745414593951; G3C5 = 5.28611102633747;

Translation of GEP5 gene into an expression tree (ET5) is shown in Fig. 4.12:

LT2C.LT2E.LT2E.GOE2B.*.OR2.Atan.X4.c7.c5.c4.d2.c4.c4.c7.c0.c4 +

X2.GOE2B.*.GOE2G.+.X2.d4.d1.d0.c7.c7.c0.c7.d2.c5.d2.d2 +

GOE2B.X4.OR2.*.GOE2B.X2.LT2C.X4.d1.c5.d5.d4.d5.d2.c9.d4.c4

GEP model-6: Classification accuracy is **100%**

y = (gepAND1(((G1C2*d(I2))-gepAND1(d(I1),d(V2))),gepG OE2B(d(V1),d(I2)))*atan((d(I3) + G1C4)));

y = y + gepAND2((d(I1)*G2C4),gepGOE2C(((d(V3) + d(I2))* d(I3)),gepLT2G(d(I1),G2C1)));

y = y + G3C8;

where, calculated value of G1C4 = 0.715048677022614; G1C2 = 4.94369335001679; G2C4 = 9.20590838343455;
G2C1 = -7.86431470686972; G3C8 = 0.398266548661763;

Translation of GEP6 gene into an expression tree (ET6) given in Fig. 4.13:

OR2.+.LT2B.c5.GOE2A.c6.+.Atan.d4.d5.c6.c8.c9.d3.c4.c9.d1

+

AND2.LT2A.LT2C.X2.3Rt.Avg2.d2.+.d3.c8.d3.d5.d2.d2.c0.d3.c6

+

OR1.+.OR1.LT2G.GOE2A.+.+.-.c7.c9.c1.d2.d2.d0.d1.d1.d3

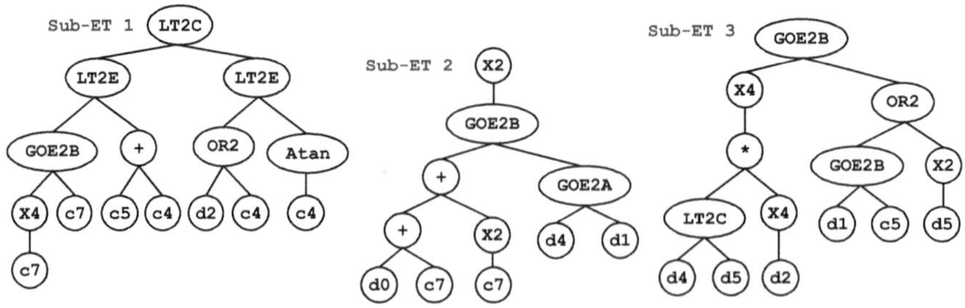

FIGURE 4.12 **ET representation for GEP model#5.**

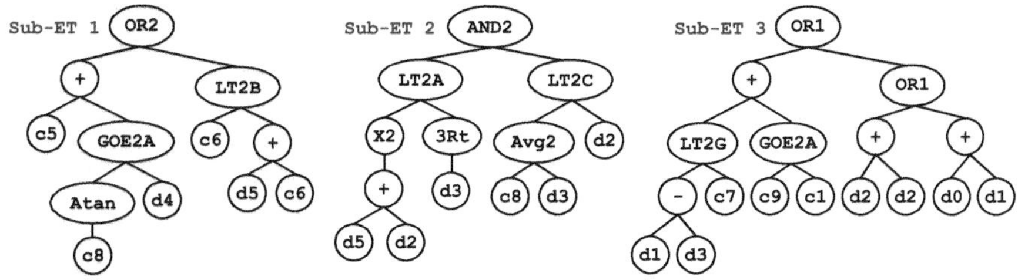

FIGURE 4.13 ET representation for GEP model#6.

The ET of GEP model-1 to GEP model-6 are demonstrated in Figs. 4.8–4.13, which represent the input relationship with respect to the output relationship of the model. The ET is represented by utilizing the mathematical equation and also represented in several computer languages as well. For the point of comparison, the MLP-ANN model and SVM models have also been designed and implemented according to the used similar data of GEP. Both ANN and SVM utilized the RMS value of the recorded signals, and diagnosis performances have been compared. The basic architecture of ANN model is 6-31-7, which is optimized by using GA optimization technique. Moreover, other operating parameters are: value of momentum factor = 0.8, tolerance = 8e-4, Mu = 1e-007, gradient = 9.2202e-6, number of epoch size = 32, rate of learning= 0.025, function at hidden layer= tangent, at output layer= linear function, MSE = 5e-3, and binary output at the output node as represented in Table 4.6. The obtained results from ANN model are represented in Fig. 4.14A,B for the training and testing phase, respectively, which show the regression value of 0.9511 and 0.88152 for training and testing phase.

Now, SVM is implemented for the demonstration of the comparison with the selected optimal value of its parameters (σ and C) of unity (1). For further validation demonstration of the performance, few cases are used as shown in Table 4.7 and its corresponding obtained results are represented in Table 4.8 by using proposed approaches, GEP as well as ANN and SVM approach.

TABLE 4.6 Network output node representation for the diagnosed condition.

Diagnosed fault	NF	OL	LR	SP	UV	OV	UB
Binary value	1000000	0100000	0010000	0001000	0000100	0000010	0000001

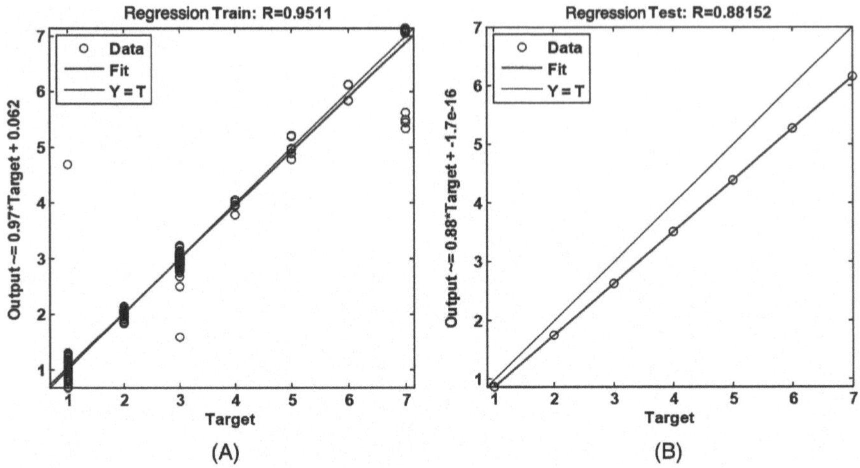

FIGURE 4.14 Graphical representation of MLP model output for (A) Training phase, (B) Testing phase.

TABLE 4.7 New samples for the further validation [3].

S. no.	Input parameters [3]						AFC
	I_1	I_2	I_3	V_1	V_2	V_3	
1	0.4277	0.4283	0.4313	2.6499	2.6398	2.6964	NF
2	0.0062	0.6447	0.6415	2.6390	2.5931	2.6670	SP
3	0.2731	0.9540	0.5640	0.9182	2.8949	2.9650	UB
4	0.3529	0.3529	0.3416	2.3824	2.3606	2.4226	UV
5	0.4835	0.4995	0.4969	2.8853	2.8800	2.8636	OV
6	1.6713	1.6505	1.6687	2.6571	2.6134	2.6873	LR
7	0.8492	0.8297	0.8431	2.642	2.6051	2.6789	OL
8	0.1756	0.5457	04814	1.3733	2.6122	2.6903	UB

AFC: actual fault condition.

From the comparative analysis of Table 4.8, it is concluded that proposed GEP approach shows excellent performance in comparison to ANN and SVM approach. Therefore, proposed approach may be used for the further future prospective.

In the further analysis of the Table 4.8 at high level of zooming, proposed approach based out results are mismatched for only few cases (sample number

TABLE 4.8 Comparative results demonstration by using ANN, SVM, and proposed GEP.

Sources of test data set	Sample no.	AFC*	Fault classification using artificial intelligence methods		
			ANN	SVM	GEP
Eight test samples taken from Ref. [3]	1	NF	NF	NF	NF
	2	SP	**NF**	SP	SP
	3	UB	UB	UB	UB
	4	UV	**UB**	UV	UV
	5	OV	OV	OV	OV
	6	LR	LR	LR	LR
	7	OL	OL	OL	OL
	8	UB	**UV**	**UV**	UB
Seventeen test samples from simulated samples in lab	1	NF	NF	NF	NF
	2	NF	NF	NF	NF
	3	SP	SP	SP	SP
	4	SP	**NF**	SP	SP
	5	UB	UB	UB	UB
	6	UB	UB	UB	UB
	7	UB	UB	UB	UB
	8	UB	UB	UB	UB
	9	UB	**SP**	UB	UB
	10	UV	UV	UV	UV
	11	UV	UV	UV	UV
	12	OV	OV	OV	OV
	13	LR	LR	LR	LR
	14	OL	OL	OL	OL
	15	OL	OL	OL	OL
	16	UB	**UV**	**UV**	UB
	17	NF	**SP**	**UB**	NF

2, 3, and 9). Moreover, ANN is unable to predict the exact diagnosis level for sample number 2, 4, and 8. Similarly, SVM model is unable to diagnose the sample number 8, 16, and 17. But for cases, proposed approach GEP is capable to diagnose all these cases according to the actual fault condition.

6 Conclusions

Fault diagnosis and condition assessment of 3-phase IM is most valuable research in the area of CM and FDD of electrical rotating machines. The MCSA is an efficient approach for diagnosing the faults of IM. This study proposed an innovative method of IM fault diagnosis, using publicly available databases of 788 patterns and 900 samples simulated in lab. For developing an accurate and consistent method of IM fault diagnosis, we have used RMS value of 3-phase voltage and current signature, defined by the Rama Hammo [3]. The proposed method for fault identification is straightforward and simple to apply in only in six steps. Additionally, proposed approach is comprehensive one as it provides diagnosis for each and every fault type. The comparison with the ANN and SVM model represents the superiority of the proposed approach. Moreover, such work would emphasis on the implementation of proposed approach for online FDCA of 3-phase induction motors.

References

[1] S.R. Kolla, L. Varatharasa, Identifying three-phase induction motor faults using artificial neural networks, ISA Trans. 39 (2000) 433–439, doi: 10.1016/S0019-0578(00)00031-8.

[2] S.R. Kolla, S.D. Altman, Artificial neural network based fault identification scheme implementation for a three-phase induction motor, ISA Trans. 46 (2007) 261–266, doi: 10.1016/j.isatra.2006.08.002.

[3] R. Hammo, Faults Identification in Three-Phase Induction Motors Using Support Vector Machines. Master of Technology Management Plan II Graduate Projects, Paper 1, BGS University, 2014. http://scholarworks.bgsu.edu/cgi/viewcontent.cgi?article=1000&context=ms_tech_mngmt. Accessed 22.06.2014

[4] J.L. Kohler, J. Sottile, F.C. Trutt, Condition monitoring of stator windings in induction motors: Part I-experimental investigation of the effective negative-sequence impedance detector, IEEE Trans. Ind. Appl. 38 (2002) 1447–1453, doi: 10.1109/TIA.2002.802935.

[5] J. Sottile, F.C. Trutt, J.L. Kohler, Condition monitoring of stator windings in induction motors: Part II-experimental investigation of voltage mismatch detectors, IEEE Trans. Ind. Appl. 38 (2002) 1454–1459, doi: 10.1109/TIA.2002.802921.

[6] A. Bellini, F. Filippetti, G. Franceschini, C. Tassoni, R. Passaglia, M. Saottini, et al., On-field experience with online diagnosis of large induction motors cage failures using MCSA, IEEE Trans. Ind. Appl. 38 (2002) 1045–1053, doi: 10.1109/TIA.2002.800591.

[7] J.P.G. De Abreu, A. Emanuel, Induction motor thermal aging caused by voltage distortion and imbalance: Loss of useful life and its estimated cost, IEEE Trans. Ind. Appl. 38 (2002) 618–624, doi: 10.1109/ICPS.2001.966519.

[8] S. Nandi, H.A. Toliya, Novel frequency-domain-based technique to detect stator inter turn faults in induction machines using stator-induced voltages after switch-off, IEEE Trans. Ind. Appl. 38 (2002) 101–109, doi: 10.1109/28.980363.

[9] R.M. Tallam, T.G. Habetler, R.G. Harley, Stator winding turn-fault detection for closed-loop induction motor drives, IEEE Trans. Ind. Appl. 39 (2003) 720–724, doi: 10.1109/TIA.2003.811784.

[10] S. Wu, T.W.S. Chow, Induction machine fault detection using SOM-based RBF neural networks, IEEE Trans. Ind. Electron. 51 (2004) 183–194, doi: 10.1109/TIE.2003.821897.

[11] H. Henao, C. Demian, G.A. Capolino, A frequency-domain detection of stator winding faults in induction machines using an external flux sensor, IEEE Trans. Ind. Appl. 39 (2004) 1272–1279, doi: 10.1109/TIA.2003.816531.

[12] S.B. Lee, G.B. Kliman, M.R. Shah, T. Mall, N.K. Nair, R.M. Lusted, An advanced technique for detecting inter-laminar stator core faults in large electric machines, IEEE Trans. Ind. Appl. 41 (2005) 1185–1193, doi: 10.1109/TIA.2005.853383.

[13] S. Nandi, Detection of stator faults in induction machines using residual saturation harmonics, IEEE Trans. Ind. Appl. 42 (2006) 1201–1208, doi: 10.1109/TIA.2006.880838.

[14] R.M. Tallam, S.B. Lee, G.C. Stone, G.B. Kliman, J. Yoo, T.G. Habetler, et al., A survey of methods for detection of stator-related faults in induction machines, IEEE Trans. Ind. Appl. 43 (2007) 920–933, doi: 10.1109/TIA.2007.900448.

[15] M. Blodt, M. Chabert, J. Regnier, J. Faucher, Mechanical load fault detection in induction motors by stator current time-frequency analysis, IEEE Trans. Ind. Appl. 42 (2007) 1454–1463, doi: 10.1109/TIA.2006.882631.

[16] C. Kral, F. Pirker, G. Pascoli, Detection of rotor faults in squirrel-cage induction machines at standstill for batch tests by means of the Vienna monitoring method, IEEE Trans. Ind. Appl. 38 (2002) 618–624, doi: 10.1109/TIA.2002.1003409.

[17] T.W.S. Chow, S. Hai, Induction machine fault diagnostic analysis with wavelet technique, IEEE Trans. Ind. Electron. 51 (2004) 558–565, doi: 10.1109/TIE.2004.825325.

[18] M. Eltabach, A. Charara, I. Zein, A comparison of external and internal methods of signal spectral analysis for broken rotor bars detection in induction motors, IEEE Trans. Ind. Electron. 51 (2004) 107–121, doi: 10.1109/TIE.2003.822083.

[19] J. Liu, P. Pillay, H. Douglas, Wavelet modeling of motor drives applied to the calculation of motor terminal over voltages, IEEE Trans. Ind. Electron. 51 (2004) 61–66, doi: 10.1109/TIE.2003.822085.

[20] C. Kral, T.G. Habetler, R.G. Harley, Detection of mechanical imbalances of induction machines without spectral analysis of time-domain signals, IEEE Trans. Ind. Appl. 40 (2004) 1101–1106, doi: 10.1109/TIA.2004.830762.

[21] B. Mirafzal, N.A.O. Demerdash, Induction machine broken-bar fault diagnosis using the rotor magnetic field space-vector orientation, IEEE Trans. Ind. Appl. 40 (2004) 534–542, doi: 10.1109/TIA.2004.824433.

[22] H. Meshgin-Kelk, J. Milimonfared, H.A. Toliyat, Interbar currents and axial fluxes in healthy and faulty induction motors, IEEE Trans. Ind. Appl. 40 (2004) 128–134, doi: 10.1109/TIA.2003.821792.

[23] C.C.M. Cunha, R.O.C. Lyra, B.C. Filho, Simulation and analysis of induction machines with rotor asymmetries, IEEE Trans. Ind. Appl. 41 (2005) 18–24, doi: 10.1109/IAS.2003.1257638.

[24] S. Bachir, S. Tnani, J.C. Trigeassou, G. Champenois, Diagnosis by parameter estimation of stator and rotor faults occurring in induction machines, IEEE Trans. Ind. Electron. 53 (2006) 963–973, doi: 10.1109/TIE.2006.874258.

[25] O. Ondel, E. Boutleux, G. Clerc, A method to detect broken bars in induction machine using pattern recognition techniques, IEEE Trans. Ind. Appl. 42 (2006) 916–923, doi: 10.1109/TIA.2006.876071.

[26] P. Neti, S. Nandi, Stator interturn fault detection of synchronous machines using field current and rotor search-coil voltage signature analysis, IEEE Trans. Ind. Appl. 45 (2009) 911–920, doi: 10.1109/TIA.2009.2018905.

[27] S. Jeevanand, A.T. Mathew, Condition monitoring of induction motors using combined PSD based wavelet decomposition and selective weighting using spider web plots, in: IEEE

Canada Electric Power Conference, Vancouver BC Canada, 2008, pp. 1–6. doi: 10.1109/EPC.2008.4763321.

[28] J.R. Stack, R.G. Harley, T.G. Habetler, An amplitude modulation detector for fault diagnosis in rolling element bearings, IEEE Trans. Ind. Electron. 51 (2004) 1097–1102, doi: 10.1109/TIE.2004.834971.

[29] J.R. Stack, R.G. Harley, T.G. Habetler, Fault-signature modeling and detection of inner-race bearing faults, IEEE Trans. Ind. Appl. 42 (2006) 61–68, doi: 10.1109/TIA.2005.861365.

[30] W. Zhou, T.G. Habetler, R.G. Harley, Bearing fault detection via stator current noise cancellation and statistical control, IEEE Trans. Ind. Electron. 55 (2008) 4260–4269, doi: 10.1109/TIE.2008.2005018.

[31] M. Blodt, P. Granjon, B. Raison, G. Rostaing, Models for bearing damage detection in induction motors using stator current monitoring, Trans. Ind. Electron. 55 (2008) 1813–1822, doi: 10.1109/TIE.2008.917108.

[32] M.B. Gulmezoglu, S. Ergin, An approach for bearing faults detection in electrical motors, Eur. Trans. Electrical Power 17 (2007) 628–641, doi: 10.1002/etep.161.

[33] Y. Xie, C. Gu, W. Cao, Study of broken bars in three-phase squirrel-cage induction motors at standstill, Int. Trans. Electrical Energy Syst. 23 (2013) 1124–1138, doi: 10.1002/etep.1642.

[34] B.M. Ebrahim, J. Faiz, B. Akin, Pattern recognition for broken bars fault diagnosis in induction motors under various supply conditions, Eur. Trans. Electrical Power 22 (2012) 1176–1190, doi: 10.1002/etep.640.

[35] M.O. Mustafa, G. Nikolakopoulos, T. Gustafsson, Broken bars fault diagnosis based on uncertainty bounds violation for three-phase induction motors, Int. Trans. Electrical Energy Syst. 25 (2015) 304–325, doi: 10.1002/etep.1843.

[36] G. Didier, H. Razik, A. Rezzoug, An induction motor model including the first space harmonics for broken rotor bar diagnosis, Eur. Trans. Electrical Power 15 (2005) 229–243, doi: 10.1002/etep.55.

[37] T. Bradatsch, D. Filbert, A parameter identification based diagnosis for detecting static air gap eccentricity, Eur. Trans. Electrical Power 5 (1995) 273–278, doi: 10.1002/etep.4450050407.

[38] M. Ojaghi, M. Sabouri, J. Faiz, Diagnosis methods for stator winding faults in three-phase squirrel-cage induction motors, Int. Trans. Electrical Energy Syst. 24 (2014) 891–912, doi: 10.1002/etep.1750.

[39] Seera M., Lim C.P., Nahavandi S., Loo C.K., Condition monitoring of induction motors: A review and an application of an ensemble of hybrid intelligent models, Expert Syst. Appl. 41 (2014) 4891–4903.

[40] Lashkari N., Poshtan J., Azgomi H.F., Simulative and experimental investigation on stator winding turn and unbalanced supply voltage fault diagnosis in induction motors using Artificial Neural Networks, ISA Trans. 59 (2015) 334–342.

[41] R.H.C. Palácios, et al., A comprehensive evaluation of intelligent classifiers for fault identification in three-phase induction motors, Electric Power Syst. Res. 127 (2015) 249–258.

[42] Sharmaa S., Malik H., Khatri A., External fault classification experienced by three-phase induction motor based on multi-class ELM, Procedia Comput. Sci. 70 (2015) 814–820.

[43] Konar P., Chattopadhyay P., Multi-class fault diagnosis of induction motor using Hilbertand wavelet transform, Appl. Soft Comput. 30 (2015) 341–352.

[44] A.P. Mittal, H. Malik, S. Rastogi, V. Talur, External fault identification experienced by 3-phase induction motor using PSVM, in: Proceedings of IEEE Sixth International Conference of Power India (PIICON), 2014, pp. 1–6. doi:10.1109/POWERI. 2014.7117762.

[45] H. Malik, A.K. Yadav, S. Mishra, T. Mehto, Application of neuro-fuzzy scheme to investigate the winding insulation paper deterioration in oil-immersed power transformer, Int. J. Electrical Power Energy Syst. 53 (2013) 256–271, doi: 10.1016/j.ijepes.2013.04.023.

[46] H. Malik, Application of Artificial Intelligence for Incipient Faults Diagnosis and Condition Assessment of Power Transformer. M. Tech. Dissertation, National Institute of Technology Hamirpur, India, 2012.

[47] W. Chuan-Sheng, H. Li, K. Li-Shan, The automatic modeling of complex functions based on gene expression programming, Int. Conf. Mach. Learn. Cybern. 5 (2005) 2870–2873, doi: 10.1109/ICNC.2007.278.

[48] The Lens, search: Scholarly Works (7290) =Induction (motor (condition monitoring)). Available from:ccess on 30-10-20 https://www.lens.org/lens/search/scholar/list?q=Induction%20 motor%20condition%20monitoring&preview=true. Accessed 30.10.2020

[49] C. Ferreira, Gene expression programming: A new adaptive algorithm for solving problems, Complex Syst. 13 (2001) 87–129 http://arxiv.org/ftp/cs/papers/0102/0102027.pdf.

[50] H. Malik, S. Mishra, A.P. Mittal, Selection of most relevant input parameters using Waikato environment for knowledge analysis for gene expression programming based power transformer fault diagnosis, Int. J. Electric Power Compon. Syst. 42 (16) (2014) 1849–1861, doi: 10.1080/15325008.2014.956952.

Chapter 5

Intelligent Data Analytics for Power Quality Disturbance Diagnosis Using Extreme Learning Machine (ELM)

1 Introduction

With the increased demand of electricity to maintain the speed of development and growth in the technically changing world challenges the electrical industry to supply clean energy with highly oriented on load management and optimization. This leads to the best management of load, by fulfilling all the objectives, standard at the service end with understanding the demand pattern of various consumer of a power utility.

The planned efforts are needed to generalize the demand pattern of consumer to minimize the power quality (PQ) issued. Such prerequisite makes the PQ investigation as the most essential territory of research as of late. Viewpoints of PQ differ at client and utility perspective. This issue opens an assortment of fields of electrical building from high power electrical machines to control hardware and low power semiconductor gadgets, to wind up an imperative region of worry for control utilities, shoppers, industry, gear originator, and producers.

An electrical power system consists of mainly three sub-systems such as generation, transmission, and distribution. These three sub-systems are integrated together to supply uninterrupted electricity to consumer at various ends. Electrical power distribution system (EPDS) holds an essential position in entire electrical system. Generally, the performance of EPDS is affected due to several PQ issues. Some of PQ issues are sag, swell, interruption, transient/oscillation, and harmonics [1]. Due to these PQ issues, there is a variation in voltage, current, or frequency and results in electrical damage or mal-functionality of customer equipment. Moreover, distributors consider power quality in terms of system reliability. Equipment designer or manufacturers, on the other hand, consider quality for proper operation of equipment under tolerance limit, whereas customers consider good PQ as availability of electricity for continuous running of processes, operations, and businesses. The PQ is now an important

Intelligent Data-Analytics for Condition Monitoring. http://dx.doi.org/10.1016/B978-0-323-85510-5.00005-3

matter of concern for electrical companies and its stakeholders as it directly affects the operation and consumption costs. Each level has its own perception (according to suppliers, customers, equipment designer, and manufacturers) as effects are unique in nature.

To understand the current scenario of PQ monitoring in electrical power system, comprehensive literature review is needed. Distributed generation and transmission plays an important role in power systems and it seems to be reliable but it shows some technical issues. So, these conventional distribution systems are being replaced by smart grid system in various developed countries. Smart grid comprises of the digital technologies that allow two-way communication between the utility and its customers, and sensing various features along the transmission lines makes the grid smart. Power utilities, usually, faces the challenge to manage their complex networks for PQ monitoring and can take intelligent decision to differentiate useful PQ data from the raw data. There are several standards which are used for solving the PQ issues directly and/or indirectly. Some of such type standards are: IEEE-519 (standards for current, voltage harmonics), IEC-61000 (EMC emission requirements applies to electrical and electronic apparatus), IEEE-141 (standard for Electric Power Distribution for Industrial Plants), IEEE-142 (standard for grounding of industrial and commercial power systems), IEEE-446 (standard for emergency and standby power systems for industrial and commercial applications), IEEE-493 (standard for design of reliable industrial and commercial power systems), IEEE-1100 (standard for powering and grounding sensitive electronic equipment), IEEE-1159 (standard for monitoring electric power quality), IEEE-1250 (standard for service to equipment sensitive to momentary voltage disturbances), and IEEE-1346 (standard for evaluating electric power system compatibility with electronic process equipment).

To solve the PQ problems, most of industries are implementing several conventional approaches. While conventional approaches are sluggish in diagnosis, off-line monitoring, and provides the time-based maintenance scenario. Therefore, researches are moving toward the implementation of artificial intelligence (AI)/machine learning (ML) based approaches for power quality disturbance (PQD) analysis. Most of these approaches are based on: (1) wavelet transform based PQD analysis [2,3], (2) fuzzy-logic (FL) based PQD analysis [4], (3) empirical mode decomposition (EMD)/ensemble EMD based PQD analysis [5,6], (4) artificial neural network (ANN) (i.e., multilayer perceptron—MLP) based PQD analysis [7], (5) particle swarm optimization (PSO) based PQD analysis [8], (6) s-transform (ST) based PQD analysis [9], (7) support vector machine (SVM) based PQD analysis , (8) PNN (probabilistic neural network) based PQD analysis [10], (9) decision tree based PQD analysis [11], and (10) hybrid techniques for PQD analysis [12,13]. Whereas these AI/ML approaches suffer from different problems such as: (1) ANN is slow in operation and it requires large amount of dataset for better diagnosis, it also suffers with trapping in finding optimal global points, (2) FL is sluggish in classification and not prone to learn new characteristic of the new dataset,

(3) SVM suffers in finding optimal value of its parameters which are time consuming, (4) PNN has some data storage problem, it requires large amount of memory, and (5) s-transform has limitation of its clarity for classification.

These earlier-mentioned all problems can be vanquished by using ELM (extreme learning machine). The ELM has three main advantages such as: (1) high accuracy, (2) least user intervention, and (3) real-time learning (i.e., learning in seconds, milliseconds, micro-seconds). ELM does not require tuning procedure for its hidden layer and its output function is generalized SLFNs (single-hidden-layer feedforward networks).

The complete study is organized as follow: 1) the introduction of the study is represented in Section 1, which show the main problem and available solution as well as the research gap of the research domain, 2) the model formation and its description is represented in Section 2, 3) the Section 3 shows the proposed approach, which is developed in this study, which includes the feature extraction (in Section 3.1), feature selection (in Section 3.2) and the diagnosis method's implementation procedure (in Section 3.3), 4) demonstrated results and its discussion are represented in Section 4 and finally Section-5 represents the conclusion part of the study.

2 Model formation and description

For the demonstration of the proposed approach, a transmission line model (TLM) is emulated in the MATLAB platform to facilitate the interlink of other equipments of sub-system level in the power system network. A 20 km transmission line of 25 kV, 50 Hz rating is connected in between source and 3-phase non-linear load. For the demonstration point of view, the following parameters are utilized for the modeling of transmission line as given:

1. *Source side parameters:* Impedance $(z) = 5 + j0.3\Omega$, connection: $\Delta \rightarrow \Upsilon$, $R_p = 0.002 p.u.$, $L_p = 0.08 p.u.$, $V_{p-p} = 25kV$
2. *Load side parameters:* $R_s = 0.002 p.u.$, $L_s = 0.08 p.u.$, $V_{p-p} = 600V$, $P_{active} = 10kW$, $P_{reactive} = 100 p.u.$, 3-phase, capacitor bank of $(V_{Nominal} = 600V$ and $P_{Reactive} = 0.8\ MW)$

The emulated TLM is presented in Fig. 5.1, in which five (5) faults are shown. Out of five faults, three faults are toward source side (i.e., sag, swell, interruption) and two faults are toward load side (i.e., transients and harmonics). For the demonstration point of view for the proposed approach, six operating condition (healthy and five faulty) of emulated model are performed and extracted signals from six emulated modes are further utilized for further study. The capture signals during operating condition have been recorded and are presented graphically in Fig. 5.2 for each operating condition. Form Fig. 5.2, it is clearly pointed out that there is some deviation and/or excitation in the signal while operating condition during the fault with respect to the healthy operation.

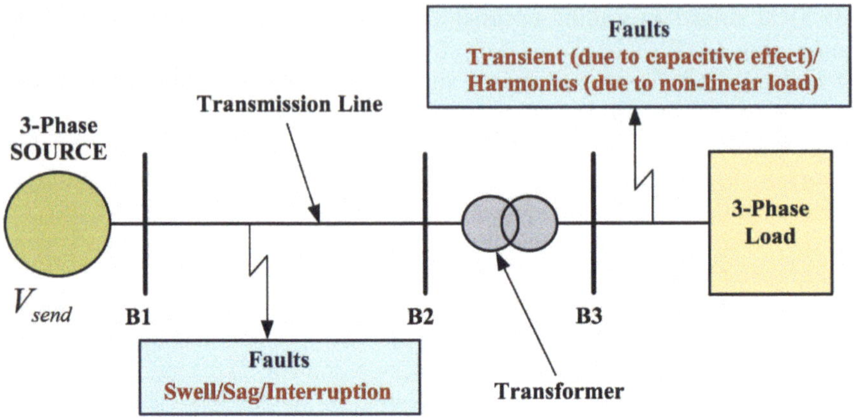

FIGURE 5.1 **Simulink model of electrical power distribution.**

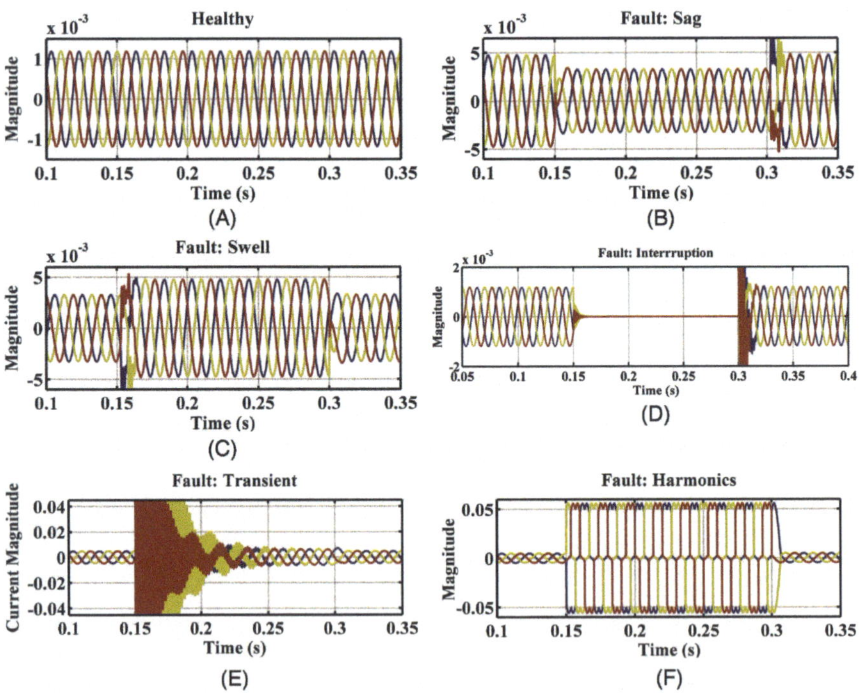

FIGURE 5.2 Current signature waveform representation for (A) Healthy condition, (B) Sag, (C) Swell, (D) Interruption, (E) Transients, and (F) Harmonics.

Therefore, these signatures may be utilized for the further study in this chapter. Figs. 5.3 and 5.4 represent the correlation of recorded current and voltage signature, respectively with respect to each other. From these figures, it is clearly concluded that recorded signals are highly correlated with respect to each other

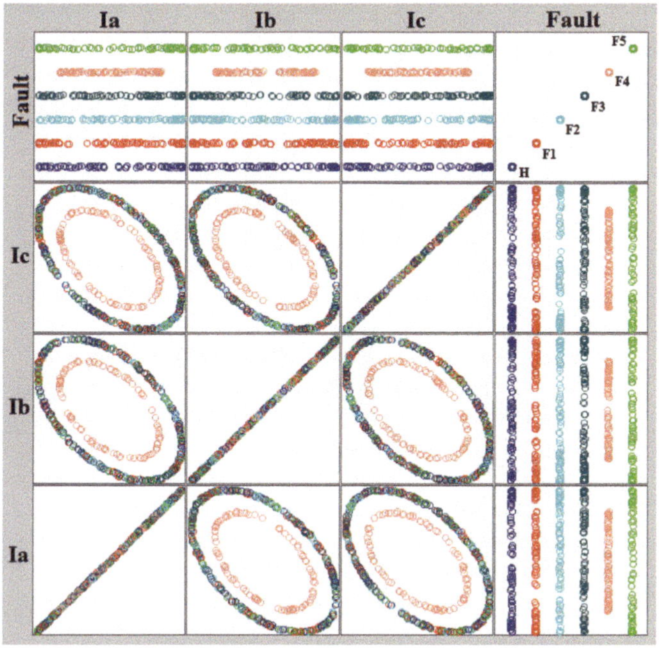

FIGURE 5.3 Visualization of recorded current dataset.

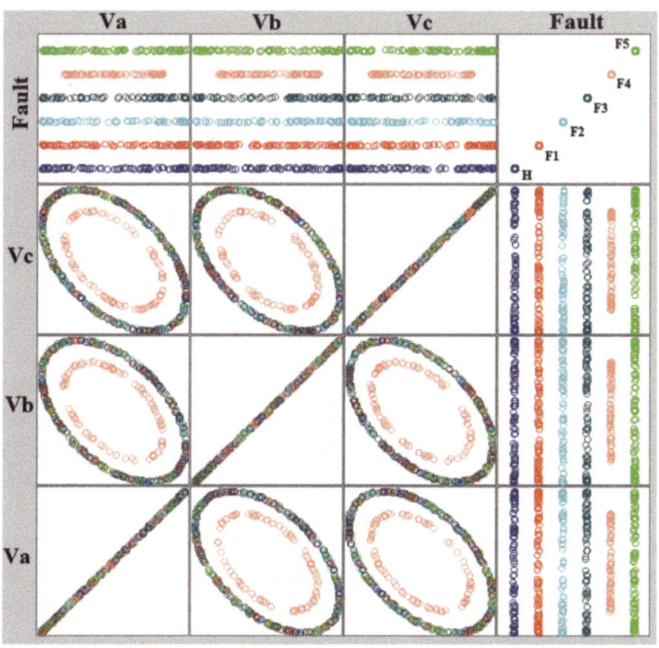

FIGURE 5.4 Visualization of recorded voltage dataset.

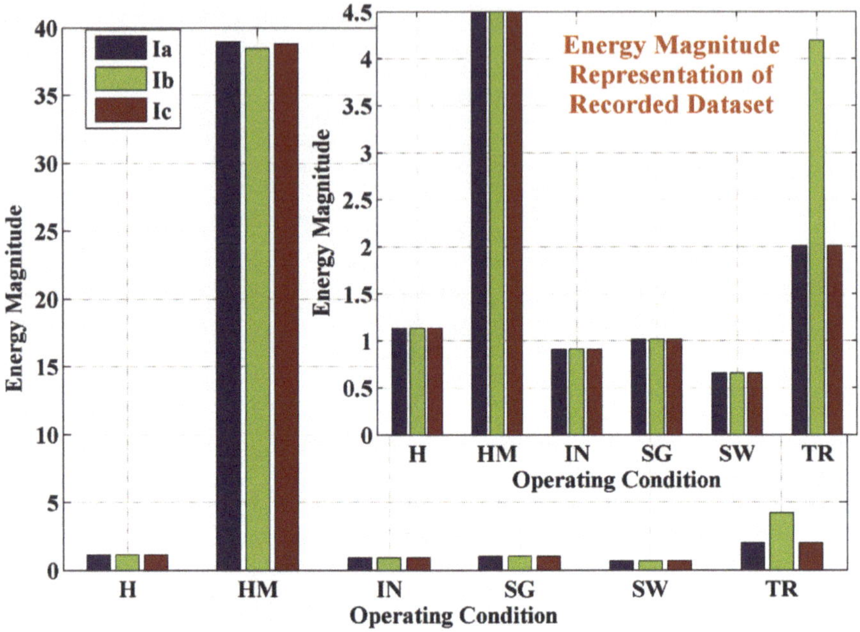

FIGURE 5.5 **Energy magnitude representation of recorded current for six different conditions.**

under different operating condition. Finally, Fig. 5.5 in this section presents the energy evaluation of the recorded signature for each operating condition along with the healthy operation. For the Fig. 5.5, it is concluded that harmonics create a lot of excitation in the system so this may be more dangerous type fault for the transmission line. Apart of harmonics faults, other faults also excited the extra energy consumption, which are comparatively higher than the healthy condition. So, the key take away point from the Fig. 5.5 is that to identify the behavior of the signal so that these signals may be utilized for further study.

3 Proposed approach

The proposed approach for the power quality disturbance diagnosis (PQDD) has been represented in Fig. 5.6, which comprises into six (6) sub-parts. These sub-parts are: (1) Part-1: model formulation, (2) Part-2: dataset collection, (3) Part-3: feature extraction, (4) Part-4: feature selection, (5) Part-5: data preparation, and (6) Part-6: intelligent classifier model development. In Part-1, transmission line model is emulated in the MATLAB environment for further uses and connected with all necessary electrical equipments. In Part-2, developed model is executed under six different operating condition for generating the data of electrical power, current, and voltage of instantaneous condition. The data generation is performed for each faulty case as well as healthy scenario.

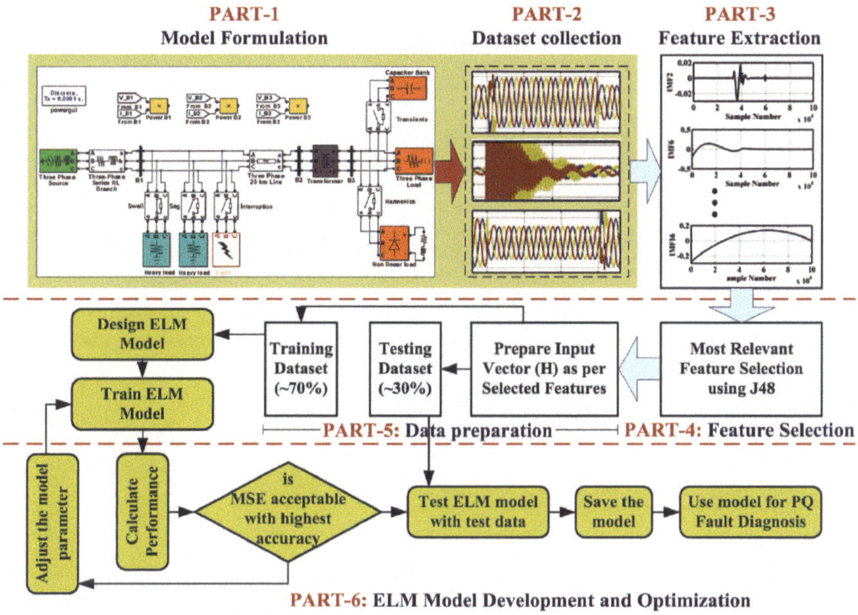

FIGURE 5.6 Proposed approach for PQ diagnosis.

Then generated dataset for each operating condition, are utilized for the further study. In Part-3, the feature extraction process is implemented. Here, EMD (empirical mode decomposition) technique for data decomposition has been implemented, which generate the number of IMFs (intrinsic mode functions), which are utilized as an input parameters to the classifiers. So, here IMFs are the extracted feature for the further analysis. After feature extraction, feature selection procedure is followed in the Part-4. In this study, a decision tree based J48 algorithm is implemented based on WEKA machine learning platform (open access platform). Feature extraction is an important part of the proposed framework because redundant features from the extracted feature matrix may be removed so that computation burden may be decrease and diagnosis accuracy will be increased, which are most important key parameters for any online implementation in real world application. After the feature extraction, redundant variables are removed then training and testing data matrixes are prepared for the model implementation in Part-5. Finally, in Part-6, ELM model is designed and then trained by using prepared training data matrix. After proper training, testing procedure is followed. And then based on testing phase performance accuracy, developed model is used for further application in PQDD.

3.1 Feature extraction using EMD technique

EMD technique is an advanced data decomposition technique of time-frequency-domain similar to wavelet transform. It is used generally to decompose

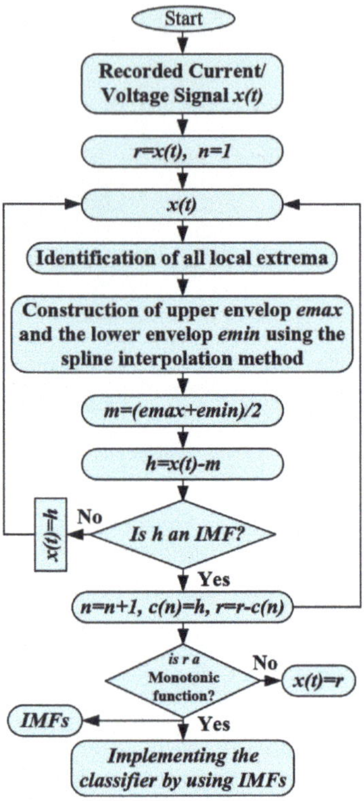

FIGURE 5.7 **Implementation procedure for EMD technique.**

non-stationary signals to generate the IMFs. The generation of IMFs depends on the two main properties. These two properties are: (1) the total number of local extrema, local minima & local maxima, and total number of zero crossing not differ by unity, (2) mean value of upper and lower envelop is zero. Based on these two conditions, EMD technique is implemented as per the shown procedure in Fig. 5.7. In this case study, 16 IMFs are generated for recorded dataset of each case.

For the better understanding about the difference in between generated all 16 IMFs of each case, the energy magnitude for each IMF of each case are evaluated are presented graphically in Figs. 5.8–5.13 for healthy, transient, swell, sag, interruption, and harmonics, respectively. After analyzing all these figures, it is concluded that there is an energy difference between each IMFs for each faulty condition as well as healthy condition. Moreover, highest excitation in the energy magnitude of the recorded signals in harmonics case is highest due to fault. And the lowest energy magnitude for recorded signal is lowest for the healthy case, which shows the energy difference between each type of recorded signal. Hence, these extracted features are used for the further analysis in this chapter. For the detailed mathematical modeling of the EMD technique, reader may refer to Refs. [14–19].

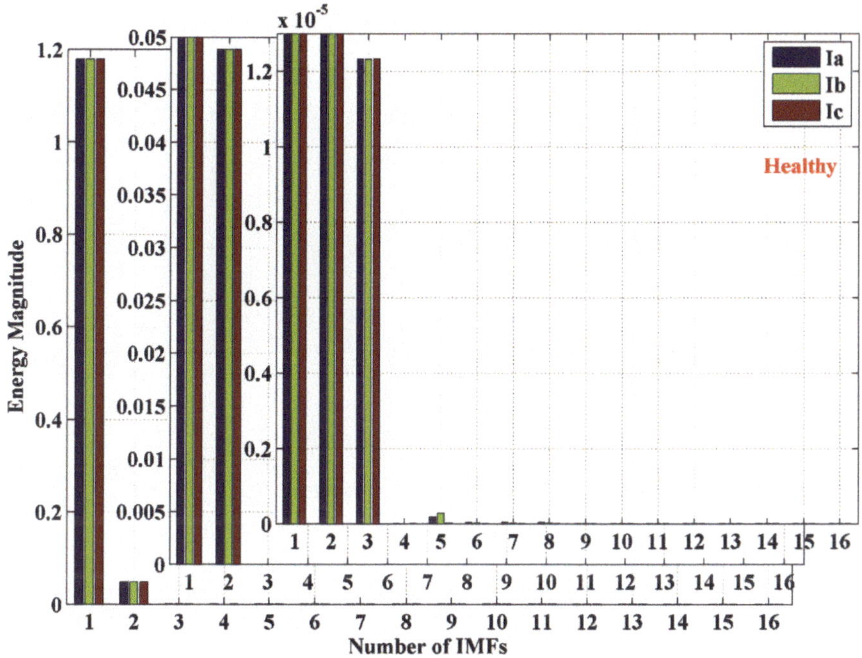

FIGURE 5.8 **Energy magnitude representation for healthy operating case.**

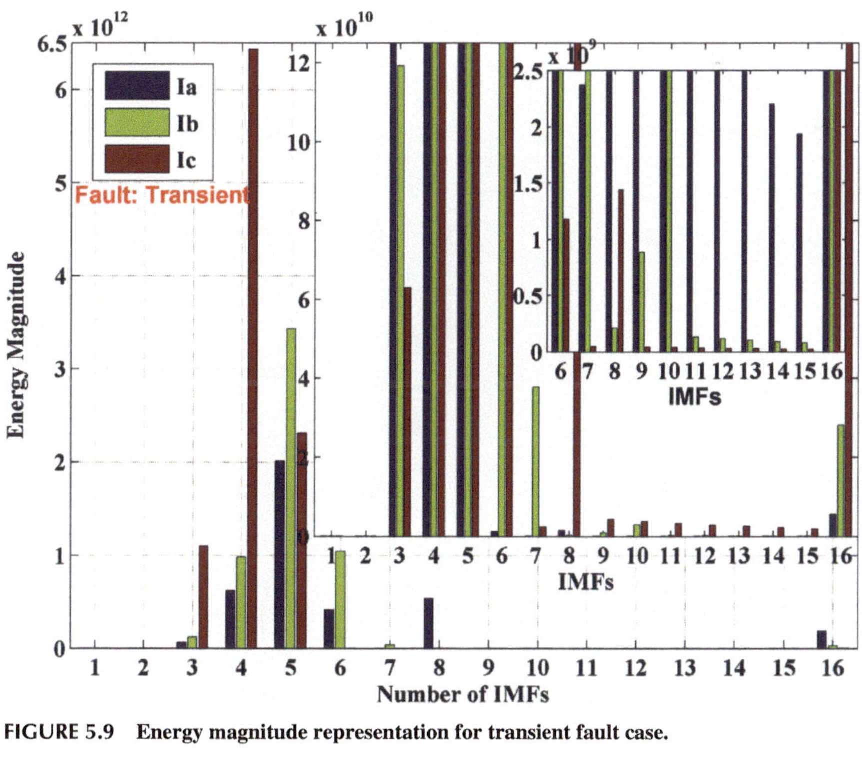

FIGURE 5.9 **Energy magnitude representation for transient fault case.**

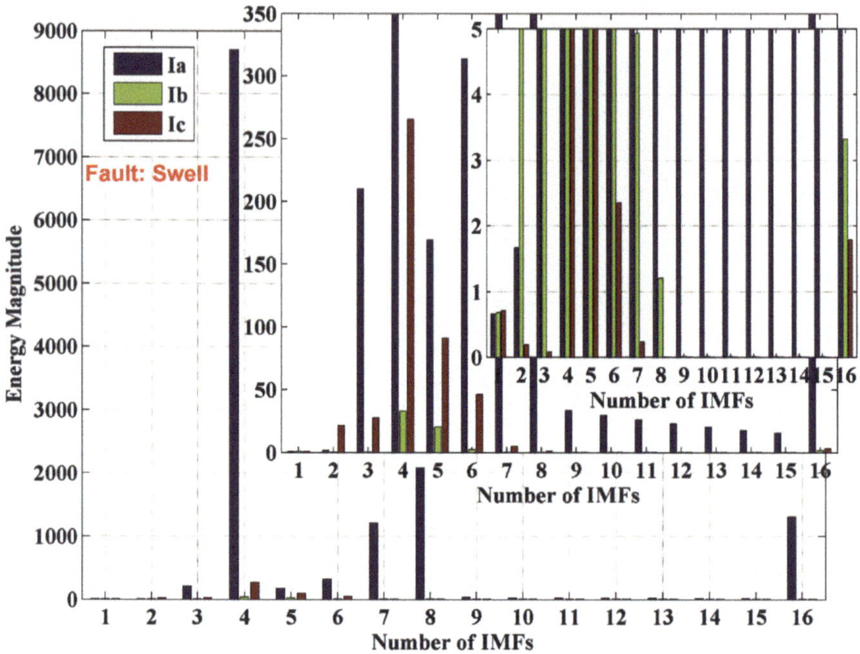

FIGURE 5.10 Energy magnitude representation for swell fault case.

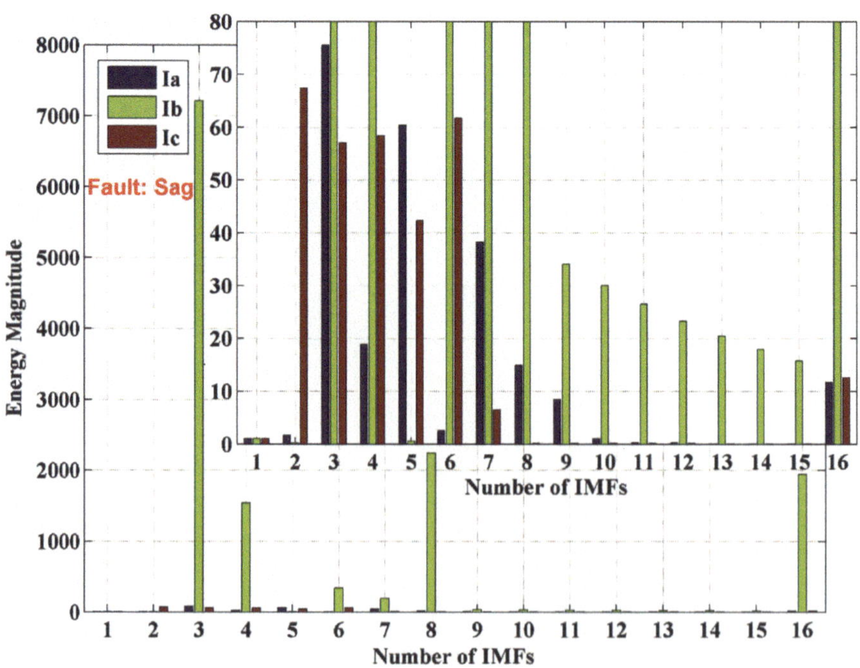

FIGURE 5.11 Energy magnitude representation for sag fault case.

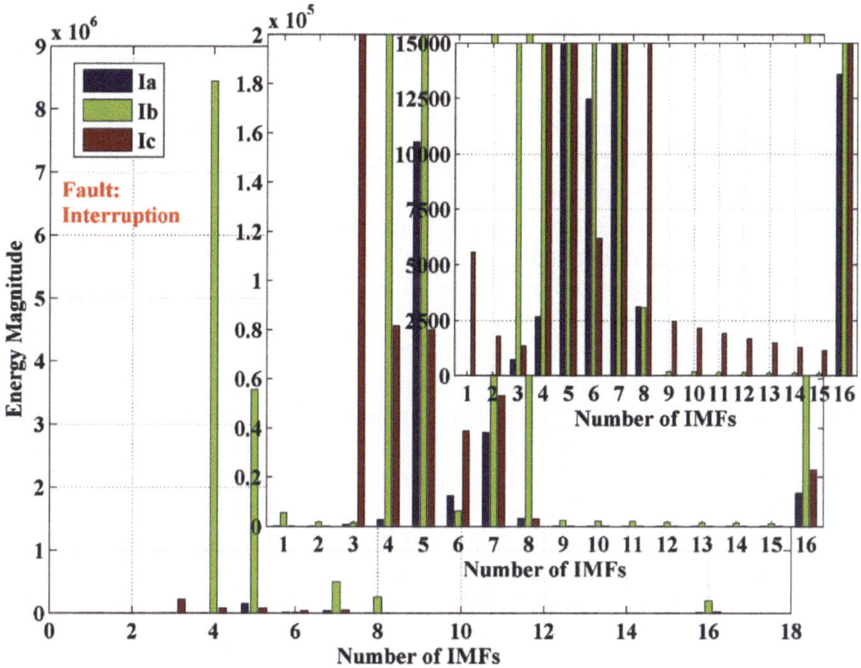

FIGURE 5.12 Energy magnitude representation for interruption fault case.

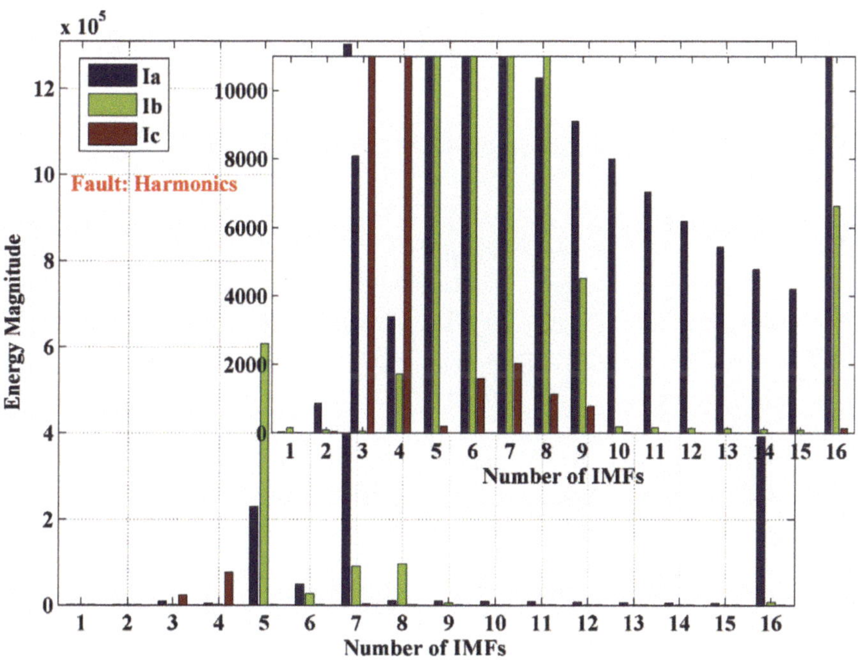

FIGURE 5.13 Energy magnitude representation for harmonics fault case.

3.2 Most relevant feature selection using WEKA based decision tree

In this section, most relevant features are selected using WEKA based J48 algorithm, which is an algorithm of decision tree. The most relevant input variable section is a process to reduce the redundant features/attributes from the feature vector so that PQDD accuracy of the model may be achieved as per satisfactory level as well as the computational burden can be reduced for the point of online implementation. In this chapter, decision tree-based J48 algorithm for feature selection has been implemented to identify the best one solution for PQDD. The detailed explanation and mathematical implementation for J48 algorithm is represented in Refs. [19–23].

The J48 algorithm is implemented first by using WEKA based machine learning platform and obtained results are represented in Table 5.1, which show the performance of the J48 algorithm's models. Here, total six models (three models based on current and three models based on voltage signals) are developed using J48 algorithm for feature selection. Developed model based on IMFs of current signals gives the best performance for PQDD, this model is chosen for the feature selection in this domain. For the detailed analysis of the J40 algorithm, all the other performance parameters of the J48 algorithm have been represented in Tables 5.2–5.4 and its graphical representation of the decision tree is shown Fig. 5.14. The nodes of the Fig. 5.14 represent the selected attributes. The main beautify of the J48 algorithm is that it removes the all redundant variable directly and presents the most relevant variables in form of its node.

TABLE 5.1 Performance analysis of PQDD interpretation for selection of most relevant input variables.

	Used data for model development	Accuracy	Kappa statistic	MAE	RMSE	Number of leave	Size of tree	Processing time (s)
Raw data based analysis	Current	31.82	0.1818	0.2223	0.3334	7	13	0.51
	Voltage	31.7267	0.1807	0.2223	0.3334	7	13	0.2
	Current & voltage	31.82	0.1818	0.2223	0.3334	7	13	0.19
Extracted feature based analysis	**IMF(Current)**	**99.99**	**0.9999**	**0**	**0.0058**	**6**	**11**	**0.4**
	IMF(Voltage)	91.8367	0.902	0.0371	0.1363	9	17	0.51
	IMF(current & voltage)	99.9933	0.9999	0	0.0047	6	11	1.41

TABLE 5.2 Prune tree representation.

```
J48 pruned tree
------------------
IMF16 <= −0.02388
|  IMF9 <= −0.005306: C3 (5000.0)
|  IMF9 > −0.005306
|  |  IMF16 <= −0.21438: C4 (5000.0)
|  |  IMF16 > −0.21438: C2 (5000.0)
IMF16 > −0.02388
|  IMF4 <= −0.12446: C1 (5000.0)
|  IMF4 > −0.12446
|  |  IMF3 <= −441.69: C5 (5000.0)
|  |  IMF3 > −441.69: C0 (5000.0)
```

TABLE 5.3 Detailed accuracy for PQDD by each class.

TP rate	FP rate	Precision	Recall	F-measure	MCC	ROC area	PRC area	Class
1.000	0.000	1.000	1.000	1.000	1.000	1.000	1.000	C0
1.000	0.000	1.000	1.000	1.000	1.000	1.000	1.000	C1
1.000	0.000	1.000	1.000	1.000	1.000	1.000	1.000	C2
1.000	0.000	1.000	1.000	1.000	1.000	1.000	1.000	C3
1.000	0.000	1.000	1.000	1.000	1.000	1.000	1.000	C4
1.000	0.000	1.000	1.000	1.000	1.000	1.000	1.000	C5
1.000	0.000	1.000	1.000	1.000	1.000	1.000	1.000	Weighted average

TABLE 5.4 Confusion matrix for PQDD by each class.

a	b	c	d	e	f	<- classified as
5000	0	0	0	0	0	| a = C0
1	4999	0	0	0	0	| b = C1
0	0	5000	0	0	0	| c = C2
0	0	0	5000	0	0	| d = C3
0	0	1	0	4999	0	| e = C4
1	0	0	0	0	4999	| f = C5

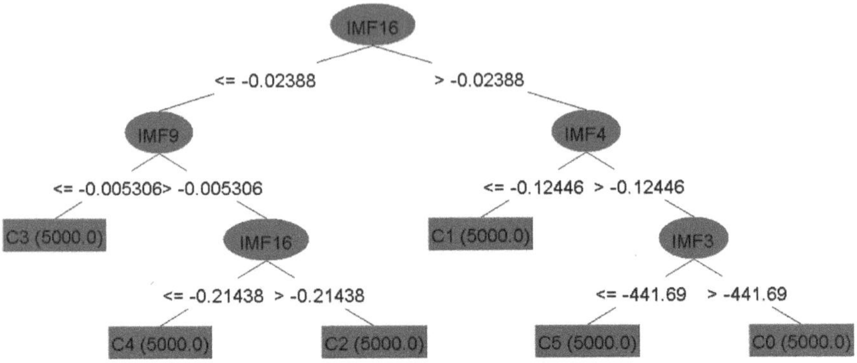

FIGURE 5.14 Decision tree of selected input variables from 16.

3.3 PQ diagnosis methods

3.3.1 *Extreme learning machine (ELM) overview*

The ELM is the FFNN (feedforward neural network), which is utilized for the data clustering, grouping, regression, prediction, forecasting, etc., with a single layer hidden node. It can generate excellent generalization performance and can learn the system 1000 of time faster than BP based NN. It is the simplest AI technique in implementation. The detailed mathematical modeling for ELM is presented in Refs. [24–28]. Moreover, the output objective function for one-hidden layer ELM is represented for Q hidden node as:

$$f_Q = \sum_{i=1}^{Q} \beta_i h_i(y) \tag{5.1}$$

Where, β_i = output weight, i = number of hidden node, h_i = output value of hidden node at i-node.

$$h_i(y) = D(a_i, b_i, y) \tag{5.2}$$

Where, a_i and b_i = parameters of hidden node.

For N number of samples, output value of hidden layer is:

$$H = \begin{bmatrix} h(y_1) \\ \bullet \\ \bullet \\ h(y_N) \end{bmatrix} = \begin{bmatrix} D(a_1, b_1, y_1) & \bullet\ \bullet\ \bullet\ \bullet & D(a_Q, b_Q, y_1) \\ \bullet & \bullet & \bullet \\ \bullet & \bullet & \bullet \\ D(a_1, b_1, y_N) & \bullet\ \bullet\ \bullet\ \bullet & D(a_Q, b_Q, y_N) \end{bmatrix} \tag{5.3}$$

Assume, target matrix used in ELM is T which includes the actual labeled value "output (o)"

$$T = \begin{bmatrix} o_1 \\ \bullet \\ \bullet \\ o_N \end{bmatrix} \tag{5.4}$$

The output weight $\beta_i : \beta_i = H^{\psi} T$.

Then the classification will be as: $Label(y_{proc}) = \text{argmax}\left[f_Q(y_{proc})\right]$.

3.3.2 Artificial neural network (ANN) overview

ANN with one hidden layer is designed as per the formulated mathematical model mentioned in Eqs. (5.5)–(5.11), which includes weight value $W^I = \left[w^I_{ih}\right]_{(n+1)\times q}$ and $W^H = \left[w^H_{hj}\right]_{(q+1)\times p}$ in between input-and-hidden layer, and between hidden-and-output layer, respectively. Fig. 5.15 encapsulates the whole simulation of MLP-ANN architecture [1,14,15]. For more information, reader may refer to Refs. [18,19].

At input Layer: the computation at the i_{th} neuron is:

$$\text{Data sample selection} = S(x_i) = x_i,\ i = 1, 2, \ldots, n; \tag{5.5}$$

$$\text{and } S(x_0) = 1. \tag{5.6}$$

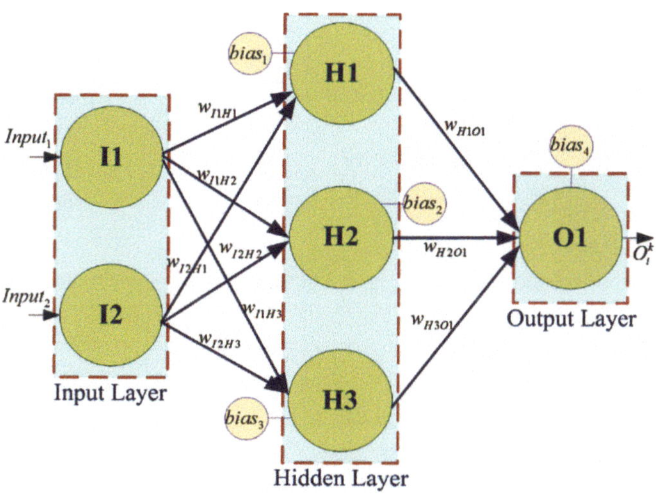

FIGURE 5.15 Single-layered MLP-ANN architecture.

At hidden Layer: The computation at the h_{th} neuron is:

$$\text{Input value } net(Hi) = z_h = \sum_{i=0}^{n} w_{ih}^{I}\, S(x_i),\ h = 1, 2, \ldots, q \tag{5.7}$$

$$\text{Output value } out\,(Hi) = S\,(z_h) = \frac{1}{1+e^{-z_h}},\ h = 1, 2, \ldots, q \tag{5.8}$$

$$S(z_0) = 1. \tag{5.9}$$

At output Layer: The computation at the j_{th} neuron is:

$$\text{Input value } out(Oi) = y_j = \sum_{h=0}^{q} w_{hj}^{H}\, S(z_h),\ j = 1,2, \ldots, p \tag{5.10}$$

$$\text{Output value } out(Oi) = S(y_j) = \frac{1}{1+e^{-y_j}},\ j = 1,2, \ldots, p \tag{5.11}$$

where, $S(\cdot)$ = activation function.

4 Results and discussion

The PQDD interpretation of MLP-ANN and ELM methods are performed to validate the proposed approach ability to diagnose the PQ issue in an effective manner. The obtained results are represented in Table 5.5. Total seven (7) models have been designed and executed for each ANN as well as ELM techniques. Table 5.5 shows that obtained results based on selected variable

TABLE 5.5 Performance analysis of ELM and MLP-ANN interpretation for PQDD.

	Dataset use	ANN based diagnosis		ELM based diagnosis	
Mode of dataset	Used data for model development	Training	Testing	Training	Testing
Raw data based analysis	Current	73.7	73.4	84.65	79.28
	Voltage	73.3	73.3	86.0	78.90
	Current & voltage	73.4	73.2	89.40	80.0
Extracted feature based analysis	IMF (current)	98.8	98.0	99.50	99.0
	IMF (current) selected	**99.0**	**98.5**	**100.0**	**99.9727**
	IMF (voltage)	97.5	96.0	98.85	97.50
	IMF (current & voltage)	98.5	93.5	99.92	98.25

excellent performance accuracy of 99% (for training), 98.5% (for testing phase) of ANN method. While training and testing accuracies for ELM case are 100% and 99.97%, respectively. Apart of this, the diagnosis accuracy based on extracted features is greater than accuracy based on raw dataset of voltage and current signature. So, these results show the advantage of EMD implementation for feature extraction for PQDD.

The performance curve for ELM and ANN models are represented in Figs. 5.16 and 5.17, respectively. Fig. 5.16 clearly shows that the performance accuracy during testing phase is varied form 98% to 100%, which is very high

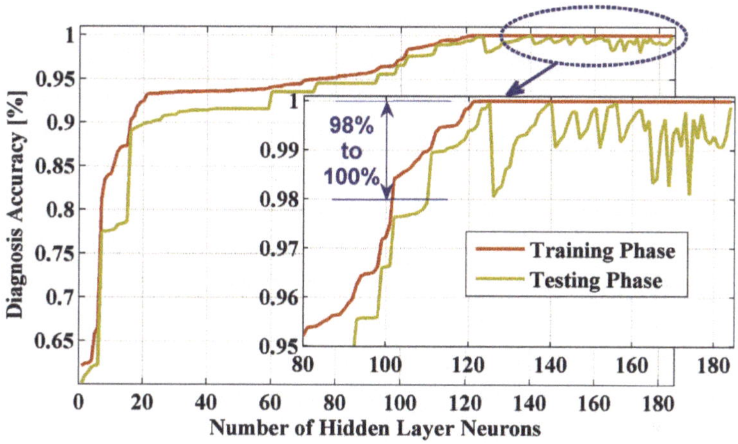

FIGURE 5.16 ELM based performance representation of proposed approach.

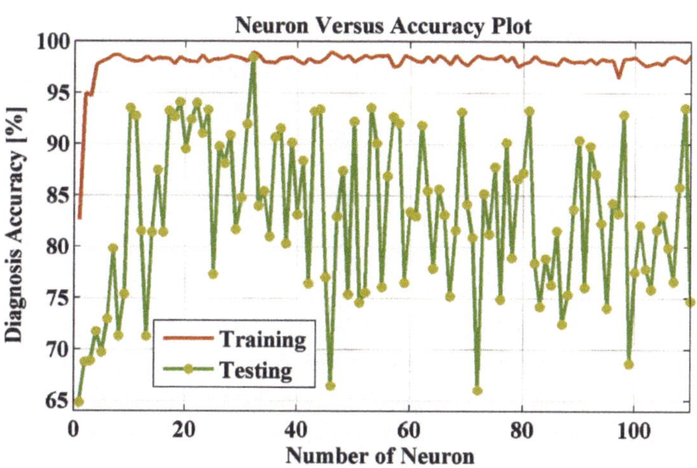

FIGURE 5.17 ANN based performance representation of proposed approach.

and acceptable. And this accuracy is achieved at only 140 hidden layer neurons, and it is under consistent manner without much deviation in variation from hidden layer neuron 120 onward.

Moreover, the performance curve for ANN is represented here in Fig. 5.17, which shows the achieved diagnosis accuracy with respect to the hidden layer neurons for the training and testing phase of the model. The training accuracy is varied upto 99% at 32 number of neurons only and the highest testing accuracy at this neuron is 98.5%, which is highly acceptable based on the proposed methodology. Apart of this, other obtained results of the model for validation point of view are represented in Figs. 5.18–5.24. Fig. 5.18 shows the performance curve

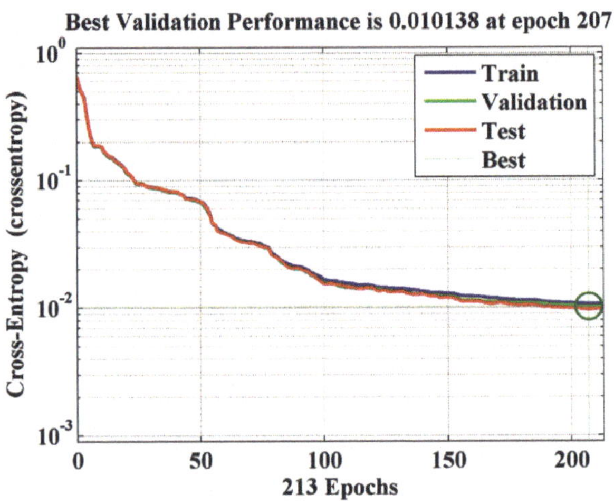

FIGURE 5.18 Performance plot of training phase.

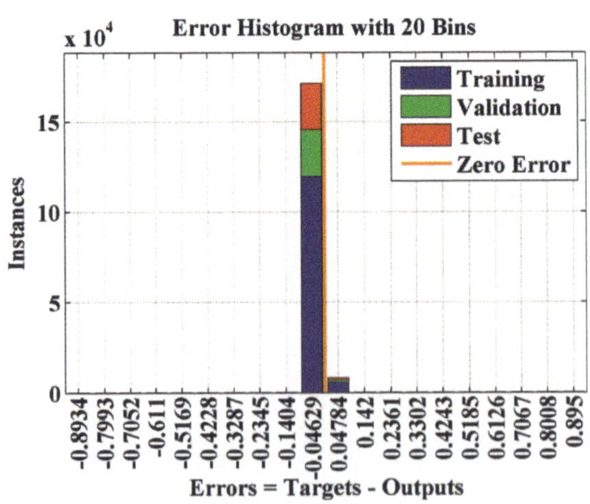

FIGURE 5.19 Error histogram plot of training phase.

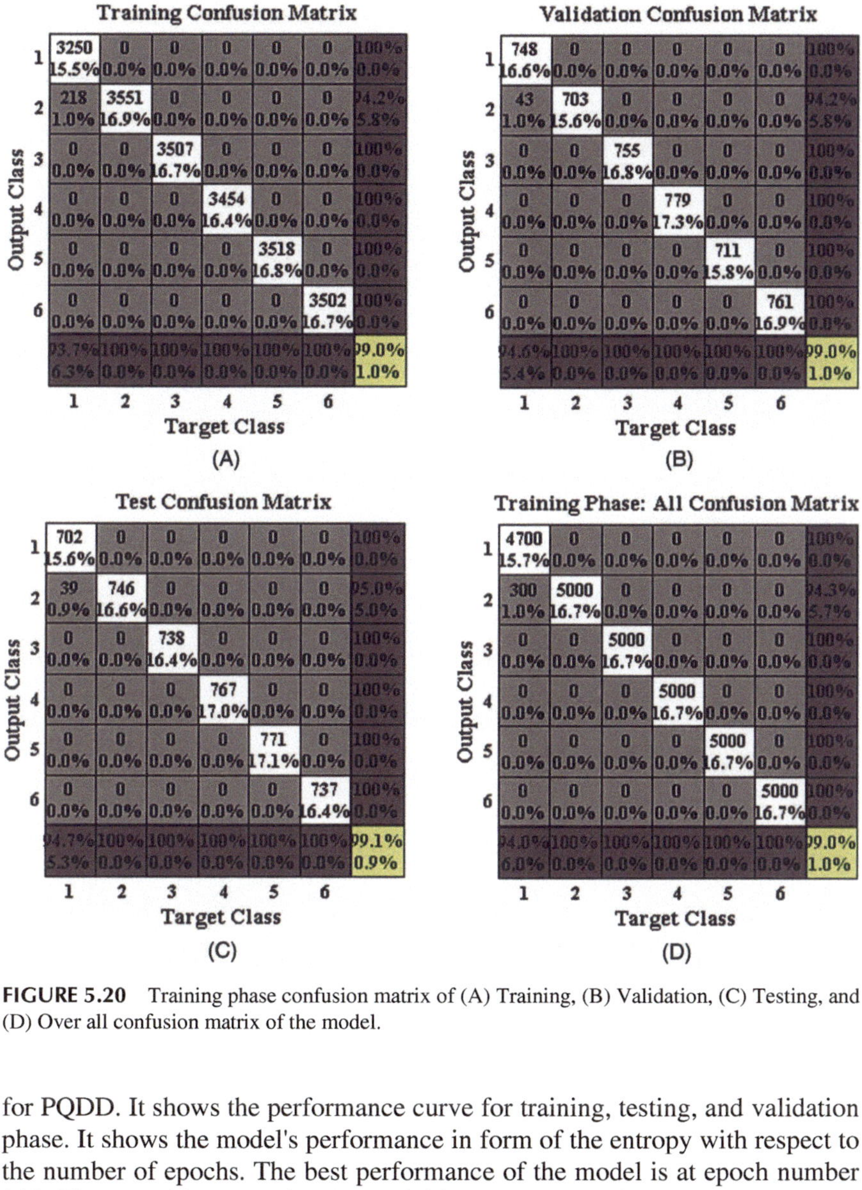

FIGURE 5.20 Training phase confusion matrix of (A) Training, (B) Validation, (C) Testing, and (D) Over all confusion matrix of the model.

for PQDD. It shows the performance curve for training, testing, and validation phase. It shows the model's performance in form of the entropy with respect to the number of epochs. The best performance of the model is at epoch number 207. The error histogram for the model is presented in Fig. 5.19, which shows the maximum and minimum range of the error value for all designing model of the ANN model. The obtained regression plots obtained from the model is shown in Fig. 5.20, which shows the correlation mapping between obtained outputs of the model with respect to the target value. Fig. 5.21 represents the ROC curve for the ANN model, which shows acceptability of the model. The training state curve of the model with respect to the gradient search method and

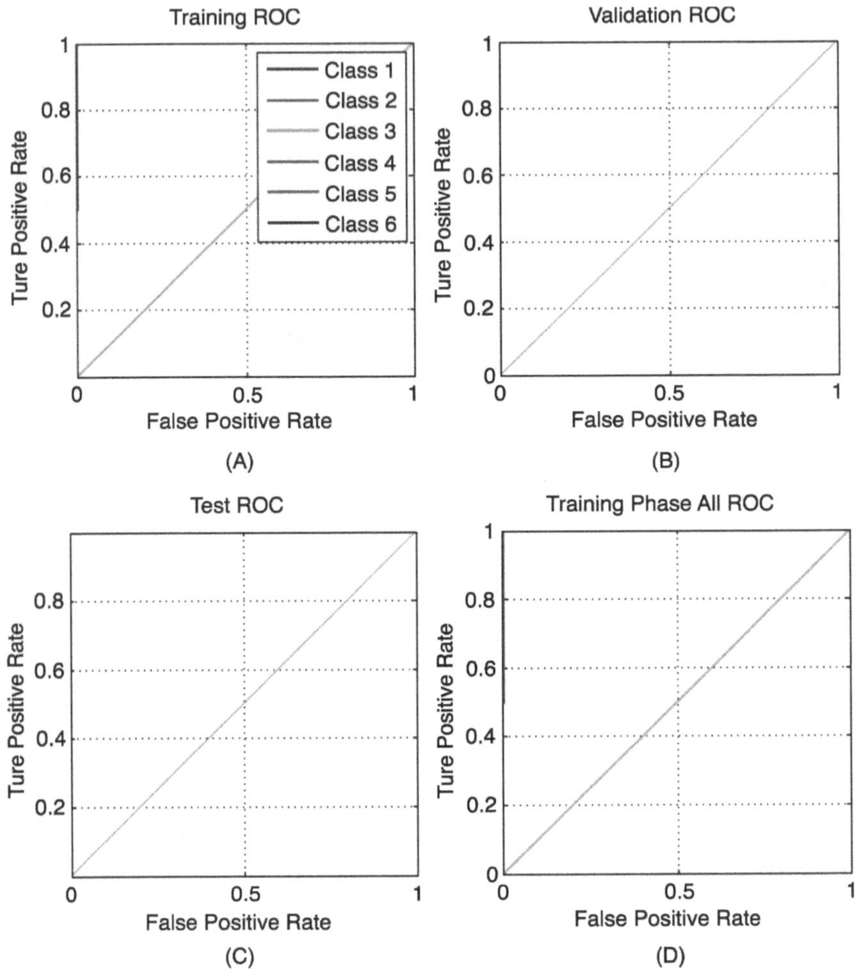

FIGURE 5.21 Training phase ROC plot of (A) Training, (B) Validation, (C) Testing, and (D) Over all ROC of the model.

validation check, is represented in Fig. 5.22, which shows the highly linear performance. Finally, testing phase results are represented in Figs. 5.23 and 5.24 for the testing phase accuracy and error histogram plot, respectively.

5 Conclusion

In this chapter, PQDD has been performed to enhance the electrical power network performance, so that power engineer may maintain the healthy operating condition to meet the uninterrupted power supply to the customer. The non-linear loading condition and other PQ phenomenon such as transient,

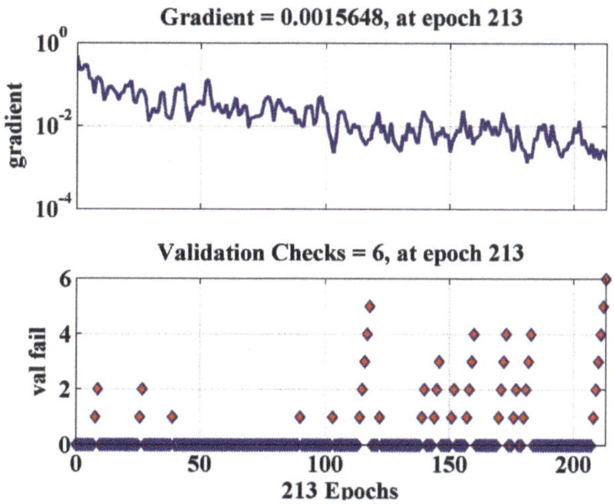

FIGURE 5.22 Training phase training status of the model.

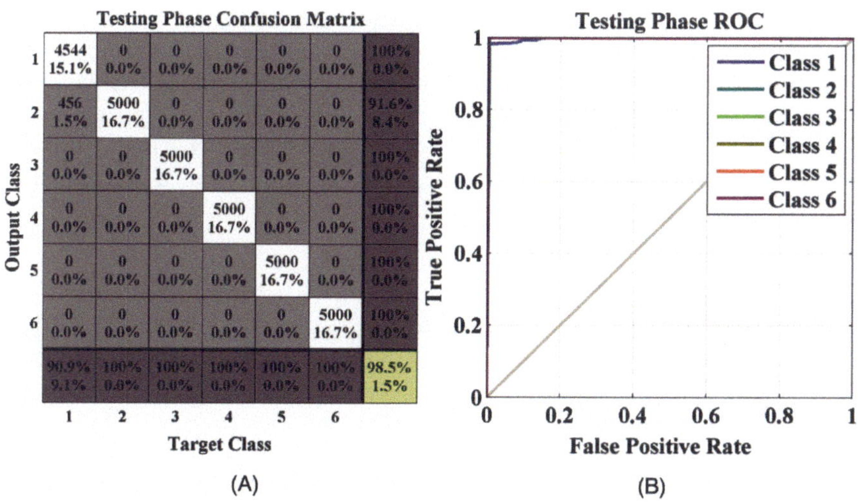

FIGURE 5.23 Testing phase plot of (A) Testing phase confusion matrix and (B) ROC curve of the model.

harmonics, sag, swell, and interruptions have been analyzed, which are the most responsible event to create the power quality disturbance problems in the power network. To enhance the reliability and stability of the power network, a novel approach for PQDD has been developed, which includes the model development, real-time data generation, data preprocessing, feature extraction, feature selection, and disturbance diagnosis. The EMD approach has been developed for feature extraction and obtained IMFs has been processed through the decision tree-based machine learning approach of J48 algorithm, which is

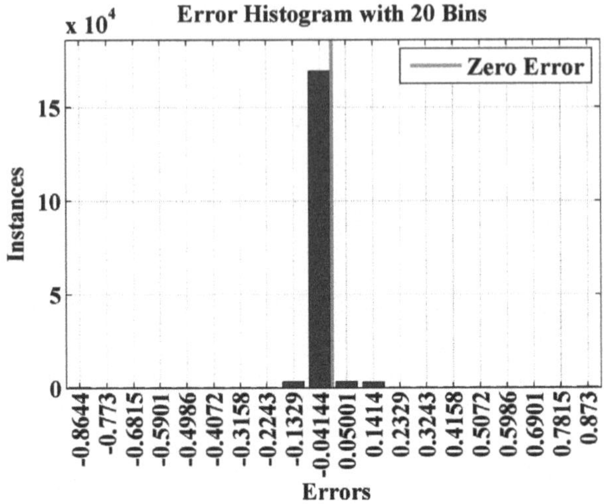

FIGURE 5.24 Testing phase plot of error histogram.

used for feature selection. In this chapter, PQDD has been performed by using ELM approach and obtained results have been compared with conventional AI method of MLP-ANN. The results showed the outperformance of the proposed approach with 99.97% for ELM and 98.5% for MLP-ANN based proposed PQDD method. The proposed approach may be implemented for real-side problem diagnosis.

References

[1] H. Malik, P. Kaushal, S. Srivastava, A hybrid intelligent model for power quality disturbance classification, Book chapter in Applications of Artificial Intelligence Techniques in Engineering, Advances in Intelligent Systems and Computing, vol. 697, Springer, Singapore, 2018, pp. 55–63, https://doi.org/10.1007/978-981-13-1822-1_6.

[2] A.M. Gaouda, E.F. El-Saadany, M.M.A. Salama, V.K. Sood, A.Y. Chikhani, Monitoring HVDC systems using wavelet multi-resolution analysis, IEEE Trans. Power Syst. 16 (4) (2001) 662–670.

[3] Jaehak Chung, J. Edward, et al., Power disturbance classifier using a rule-based and wavelet packet-based hidden Markov model, IEEE Trans. Power Deliv. 17 (1) (2002) 233–241.

[4] P.K. Dash, S. Mishra, et al., Classification of power system disturbances using a fuzzy expert system and a Fourier linear combiner, IEEE Trans. Power Deliv. 15 (2) (2000) 472–477.

[5] V. Ignatova, et al., Space vector method for voltage dips and swells analysis, IEEE Trans. Power Deliv. 24 (4) (2009) 2054–2061.

[6] Z. Li, et al., A classification method for complex power quality disturbances using EEMD and rank wavelet SVM, IEEE Trans. Smart Grid 6 (4) (2015) 1678–1685.

[7] R. Kumar, B. Singh, et al., Recognition of power-quality disturbances using s-transform-based ANN classifier and rule based decision tree, IEEE Trans. Power Deliv. 51 (2) (2015) 1249–1258.

[8] R. Hooshmand, A. Enshaee, Detection and classification of single and combined power quality disturbances using fuzzy systems oriented by particle swarm optimization algorithm, Electric Power Syst. Res. 80 (2010) 1552–1561.
[9] S. He, et al., A real-time power quality disturbances classification using hybrid method based on s-transform and dynamics, IEEE Trans. Instr. Meas. 62 (9) (2013) 2465–2475.
[10] S.R. Mohanty, et al., Classification of disturbances in hybrid DG system using modular PNN and SVM, Int. J. Electrical Power Energy Syst. 44 (2013) 764–777.
[11] P.K. Ray, et al., Optimal feature and decision tree based classification of power quality disturbances in distributed generation systems, IEEE Trans. Sustain. Energy 5 (1) (2014) 200–208.
[12] J. Huang, et al., Neural-fuzzy classifier for recognition of power quality disturbances, IEEE Trans. Power Deliv. 17 (2) (2002) 609–616.
[13] M. Sabarimalai Manikandan, et al., Detection and classification of power quality disturbances using sparse signal decomposition on hybrid dictionaries, IEEE Trans. Instr. Meas. 64 (1) (2015) 27–38.
[14] H. Malik, R. Sharma, EMD and ANN based intelligent fault diagnosis model for transmission line, J. Intell. Fuzzy Syst. 32 (4) (2017) 3043–3050, doi: 10.3233/JIFS-169247.
[15] A.K. Shah, A. Yadav, H. Malik, EMD and ANN based intelligent model for bearing fault diagnosis, J. Intell. Fuzzy Syst. 35 (5) (2018) 5391–5402, doi: 10.3233/JIFS-169821.
[16] H. Malik, A. Aggarwal, R. Sharma, Feature extraction using EMD and classification through probabilistic neural network for fault diagnosis of transmission line, in: Proceedings of IEEE First International Conference on Power Electronics, Intelligent Control and Energy System (ICPEICES), Delhi, India, 2016, pp. 1–6. doi: 10.1109/ICPEICES. 2016.7853709
[17] H. Malik, D. Chack, A novel intelligent transmission line fault diagnosis model based on EEMD and multiclass PSVM, Book chapter in Applications of Artificial Intelligence Techniques in Engineering, Advances in Intelligent Systems and Computing, 698, Springer, Singapore, 2018, pp. 85–92, doi: https://doi.org/10.1007/978-981-13-1819-1_9.
[18] H. Malik, Y. Pandya, A. Prashar, R. Sharma, Feature extraction using EMD and classifier through artificial neural networks for gearbox fault diagnosis, Book chapter in Applications of Artificial Intelligence Techniques in Engineering, Advances in Intelligent Systems and Computing, 697, Springer, Singapore, 2018, pp. 309–317 https://doi.org/10.1007/978-981-13-1822-1_28.
[19] H. Malik, S. Mishra, Artificial neural network and empirical mode decomposition based imbalance fault diagnosis of wind turbine using TurbSim, FAST and Simulink, IET Renew. Power Generat. 11 (6) (2017) 889–902, doi: 10.1049/iet-rpg.2015.0382.
[20] H. Malik, S. Mishra, Application of gene expression programming (GEP) in power transformers fault diagnosis using DGA, IEEE Trans. Ind. Appl. 52 (6) (2016) 4556–4565, doi: 10.1109/TIA.2016.2598677.
[21] H. Malik, S. Mishra, Application of GEP to investigate the imbalance faults in direct-drive wind turbine using generator current signals, IET Renew. Power Generat. 12 (3) (2018) 279–291, doi: 10.1049/iet-rpg.2016.0689.
[22] H. Malik, R. Sharma, S. Mishra, Fuzzy reinforcement learning based intelligent classifier for power transformer faults, ISA Trans. 101 (2020) 390–398 https://doi.org/10.1016/j.isatra.2020.01.016.
[23] H. Malik, S. Mishra, A.P. Mittal, Selection of most relevant input parameters using Waikato environment for knowledge analysis for gene expression programming based power transformer fault diagnosis, Int. J. Electric Power Compon. Syst. 42 (16) (2014) 1849–1862, http://dx.doi.org/10.1080/15325008.2014.956952.

[24] H. Malik, S. Mishra, Selection of most relevant input parameters using principle component analysis for extreme learning machine based power transformer fault diagnosis model, Int. J. Electric Power Compon. Syst. 45 (12) (2017) 1339–1352, https://doi.org/10.1080/15325008.2017.1338794.

[25] H. Malik, S. Mishra, Extreme learning machine based fault diagnosis of power transformer using IEC TC10 and its related data, in: Proceedings of IEEE India Annual Conference (INDICON-2015), New Delhi, India, 2015, pp. 1–5, doi: 10.1109/INDICON.2015.7443245

[26] H. Malik, S. Mishra, Application of extreme learning machine (ELM) in paper insulation deterioration estimation of power transformer, in: Proceedings of the International Conference on Nanotechnology for Better Living, vol. 3 (1), 2016, pp. 209. doi: 10.3850/978-981-09-7519-7nbl16-rps-209

[27] H. Malik, N. Roy, Extreme learning machine-based image classification model using handwritten digit database, Book chapter in Applications of Artificial Intelligence Techniques in Engineering, Advances in Intelligent Systems and Computing, 697, Springer, Singapore, 2018, pp. 607–618, https://doi.org/10.1007/978-981-13-1822-1_57.

[28] S. Sharma, H. Malik, A. khatri, External fault classification experienced by three-phase induction motor based on multi-class ELM, Procedia Comput. Sci. 70 (2015) 814–820, doi: 10.1016/j.procs.2015.10.122.

Chapter 6

Intelligent Data Analytics for Transmission Line Fault Diagnosis Using EEMD-Based Multiclass SVM and PSVM

1 Introduction

The enhancing severity of the electrical power network (EPN) and maintaining uninterrupted operation presents several encounters with the fault detection and diagnosis (FDD) of the system. There are numerous reasons behind the blackouts in a particular country. Few statistics for outage cause categories (in percentage) in the United States are: earthquake (0.8%), tornado (2.8%), hurricane/tropical storm (4.2%), ice-storm (5%), lightning (11.3%), wind/rain (14.8%), other cold water (5.5), fire (5.2%), intentional attack (1.6%), supply shortage (5.3%), equipment failure (29.7%), operator error (10.1%), voltage reduction (7.7%), volunteer reduction (5.9%), and other external cause (4.8%)[a]. Note that total blackout events are more than 100% because some events occur in multiple event categories. So, the operation & maintenance (O&M) cost of transmission & distribution (T&D) and substation is increased drastically. According to Southern California Edison (SCE) report-2015[b], the O&M cost of different components are: (1) for T&D & substation 13%, (2) for grid operation 18%, (3) for pole loading 6%, (4) for distribution maintenance 30%, (5) for customer driven prog & distr con 2%, (6) for T&D engineering and grid technology 8%, (7) for T&D other cost and other operating revenue (OOR) 2%, and (8) for safety, training, and environmental programs 11%. This cost can be minimized by using proper health monitoring systems in the T&D lines.

According to Ref. [1], approximate 80% faults of electrical power network take place in overhead transmission lines, which are diagnosed into two categories such as: (1) fault classification and (2) fault location identification. Generally, the FDD approaches of the transmission line (TL) include feature

[a]https://fas.org/sgp/crs/misc/R42696.pdf.

[b]http://www3.sce.com/sscc/law/dis/dbattach5e.nsf/0/BACF374E1EFE27CE88257C2100811E3D/$FILE/SCE-03%20Vol.%2008.pdf.

Intelligent Data-Analytics for Condition Monitoring. http://dx.doi.org/10.1016/B978-0-323-85510-5.00006-5

extraction, feature selection, and feature classification procedure. The feature extraction procedure extracts the relevant and/or non-relevant information from the recorded raw signals. The feature selection procedure develops the correlation and shows the relevancy in between the extracted feature and TL AOC (actual operating condition), which is identified by using the classification/identification method. In the TL fault scenario, the useful information related to its appropriate fault is available in post fault voltage and/or current signals. So in this domain, numerous methods have been proposed by several researchers to specify the available information in the signals. Though, choosing the most suitable approach and feature depends on expert knowledge and performance. However, most relevant features can be selected by using feature selection approaches [2–6].

The input feature extraction procedure enhances the burden on the computation of the FDD process and leads the sluggish operation in the online FDD of TL. In the recent development of FDD algorithms and methods in EPN and TL has observed a similar design and development with opting different similar problem [1,7–11]. The utilization of classical method [7], wavelet [8], neuro-fuzzy [9], support vector machine (SVM) [10], fuzzy-logic [11], reinforcement learning [12], artificial neural network [13–15], decision tree (DT) [16], etc., are the initial attempts in introducing the advanced methods in EPN FDD. However, the conventional approach highly rely on expert knowledge and skills, FL-based approaches have non-dynamic adaptability feature, ANN-based approaches are sluggish in operation due to a large amount of the training dataset, SVM requires optimization of its operating parameters, DT approach depends on advanced decision logic.

Therefore, key-limitations of AI (artificial intelligence) and ML (machine learning) methods for TL FDD exist around in TL's two-categories only as mentioned earlier (i.e., (1) fault classification and (2) fault location identification). However, for analyzing these two-categories of TL, two different methods/approaches are required individually [1] because both type of methods require individual parameter's setting and training into different modes of operation. Therefore, an approach/method that can achieve high efficiency in a short period of processing time is the demand of the TL system.

Latterly, a novel EEMD (ensemble empirical mode decomposition) based on an intelligent FDD model using proximal support vector machine (PSVM) has been proposed and obtained results are compared with standard SVM, which presents superiority of the proposed approach in term of diagnostic accuracy enhancement, reduction in computational time, and programming complexity [17–20].

The chapter is organized as follows. Section 2 presents the methodology which includes the model description, brief detail of EEMD, SVM, and PSVM method. Multi-class SVM and PSVM fault diagnostic models are also depicted in Section 2. The obtained results are presented and discussed in Section 3. In Section 3, a comparative study of SVM and PSVM based fault diagnosis is also presented followed by a conclusion in Section 4.

2 Methodology

2.1 Proposed approach

The proposed approach framework for TL FDD is shown in Fig. 6.1, which include all the steps from model formation to FDD in step-wise-step manner as follows:

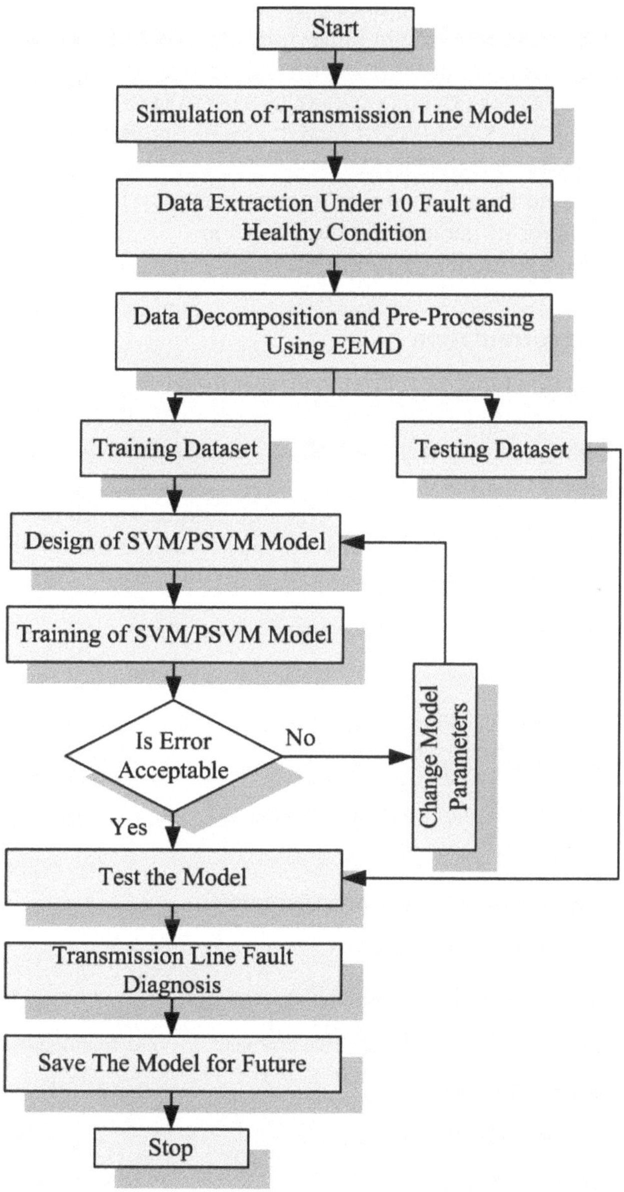

FIGURE 6.1 **Proposed approach flowchart for transmission line fault diagnosis.**

Step#1: TL model formation with healthy and 10 different fault condition
Step#2: Raw data extraction during all 11 operating condition of TL
Step#3: Data preprocessing—during preprocessing process, following operations have been performed: missing value filling, removal of spicks, normalization of the data
Step#4: Feature extraction using EEMD method
Step#5: Prepare the training data (~70%), testing data (~30%) separately and along with its target file in binary mode [10000000000; 01000000000; 00100000000; 00010000000; 00001000000; 00000100000; 00000010000; 00000001000;00000000100;00000000010;00000000001]
Step#6: Design the SVM/PSVM model with its optimal parameters
Step#8: Train the designed models
Step#9: Test the trained model if training accuracy is sufficient else train again with and/or its parameter value variation
Step#10: Save the model for future TL FDD

2.2 TL model formulation

To represent the suitability of PSVM approach for TL FDD in EPN, a 300 km, 500 kV, 50 Hz TL has been designed as shown in Fig. 6.2. This model is developed according to available real-TL with all suitable parameters as represented in Table 6.1. The emulation of model of Fig. 6.2 generates the raw datasets for all 11 cases (i.e., healthy case and 10 types of fault cases). In this study, three different categories of faults are analyzed such as single line to ground faults

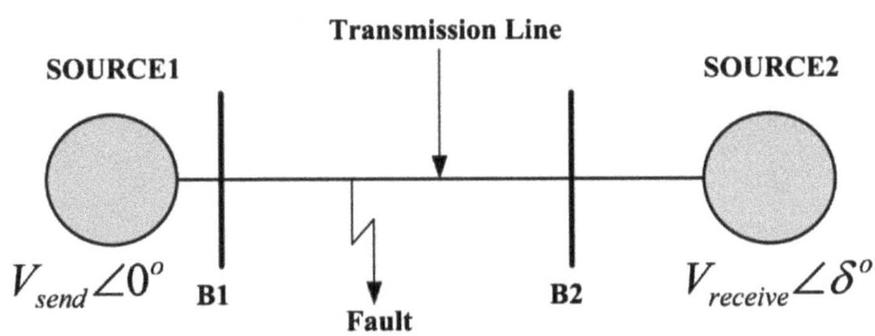

FIGURE 6.2 Simplified 300 km transmission line model.

TABLE 6.1 TL's parameters used in this study work.

Name of parameters	Sending end point	Receiving end point
Source voltage	500 kV L-L, 0°	475 kV L-L, 15°
Z_{S1}	17.177 + 45.5285 jΩ	15.31 + 45.9245 jΩ
Z_{S2}	2.5904 + 14.7328 jΩ	0.37432 + 1.1315 jΩ

(3 numbers), double-line to ground faults (3 numbers), and short circuited faults (4 numbers). The symbolic representation of each case of faults is shown as: HI-Healthy, LGAG-single line (A) to ground fault, LGBG-single line (B) to ground fault, LGCG-single line (C) to ground fault, LLGABG-double line (AB) to ground fault, LLGBCG-double line (BC) to ground fault, LLGCAG-double line (CA) to ground fault, LLAB-double line (AB) short-circuit fault, LLBC-double line (BC) short-circuit fault, LLCA-double line (CA) short-circuit fault, and LLLABC-triple line (ABC) short-circuit fault. Here, the notations A, B, and C are used for the representation of three lines of a 3-phase TL.

Transmission line parameters:

$$C_0 = 7.4982E-9F/KmC_1 = 1.2097E-8F/Km$$
$$Z_0 = 1.2097E-8F/KmZ_1 = 0.15483+0.36608\,j\Omega/Km$$

Graphical representation of recorded current for emulated 10 faults and healthy conditions is represented in Fig. 6.3.

2.3 Feature extraction using EEMD

EEMD is the advance version of EMD algorithm which is use to overcome the problem of mode mixing of EMD approach. The implementation process of EEMD algorithm comprises into three simple steps as given [21]:

Step#1: Add the enough amount of white noise $\cong$ mean and variance of original signal

Step#2: Decompose each new signal (original signal + noise) by using EMD algorithm

Step#3: Compute the average of all IMFs obtained by EMD at Step#2 to remove the effect of white noise

Fig. 6.4 represents the mathematical modeling in form of flowchart of EEMD algorithm. The addition of noise furnishes the uniform base scale distribution to reduce mode mixing for EMD based decomposition. This procedure increases the accuracy and reduces the false frequency in IMFs.

To analyze strength variation of recorded non-linear signals under 10 faulty and 1 healthy condition of transmission line, energy magnitude of generated IMFs of EEMD is presented in Fig. 6.5A–L.

2.4 Support vector machine (SVM)

SVM is principally used for finding the solutions for problems of regression and classification or can be used for outlier detection [10]. SVM is used specifically where there are two classes of datasets. SVM classifies dataset by hyperplane, which separates all points of datasets into defined separate classes. Best SVM hyperplane margin is achieved where there is a largest margin in two classes of the hyperplane and here margin is the maximum width of the slab parallel to the hyperplane which is without interior points of datasets, as shown in Fig. 6.6A.

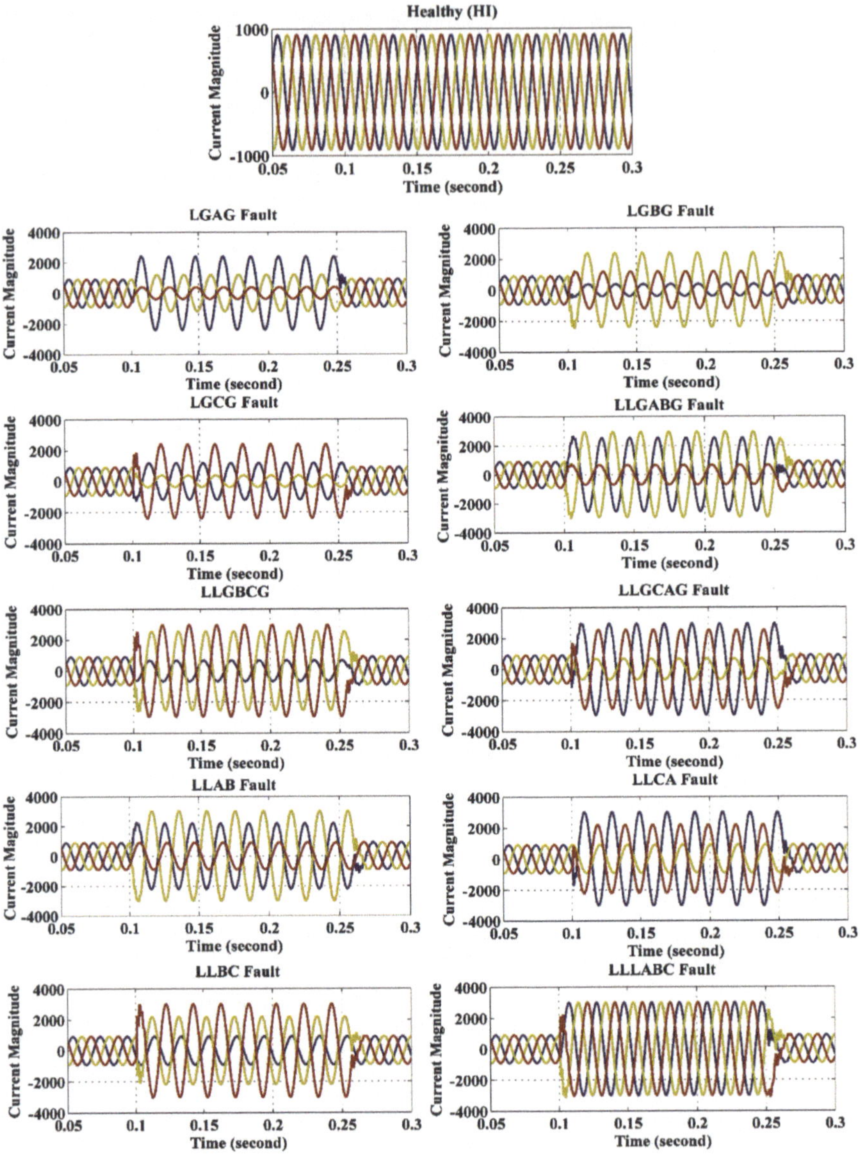

FIGURE 6.3 Graphical representation of current signal of 11 conditions (healthy condition, LAG-LBG-LCG-LABG-LBCG-LCAG-LAB-LBC-LCA-LABC).

The support vectors (SVs) are the dataset points which are nearer to the separating hyperplane, these points are on the slab boundary. Fig. 6.6A represented as +ve (positive) data points for class type#1 and –ve (negative) dataset points as class type#2. The implementation modeling of SVM classifier used in this study is represented in Eq. (6.4).

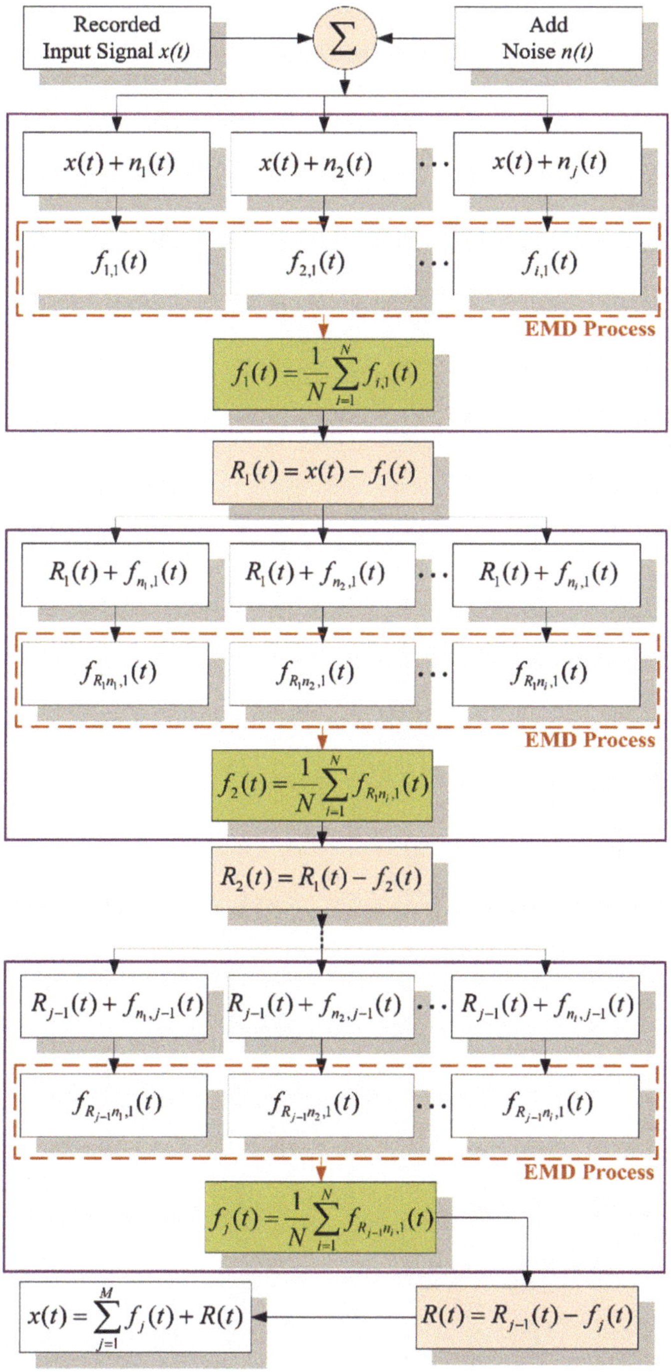

FIGURE 6.4 The EEMD algorithm's flowchart.

For the modeling of the SVM classifier, the training dataset S_j and its associated labeled class T_j are utilized. The class categories T_j is represented in the form of ±1 and the hyperplane equation can be represented in Eq. (6.1) for the dataset dimension d $\left(\therefore S_j \in R^d s\right)$ [4,7,14–16]:

$$< \omega, S > + b = 0 \tag{6.1}$$

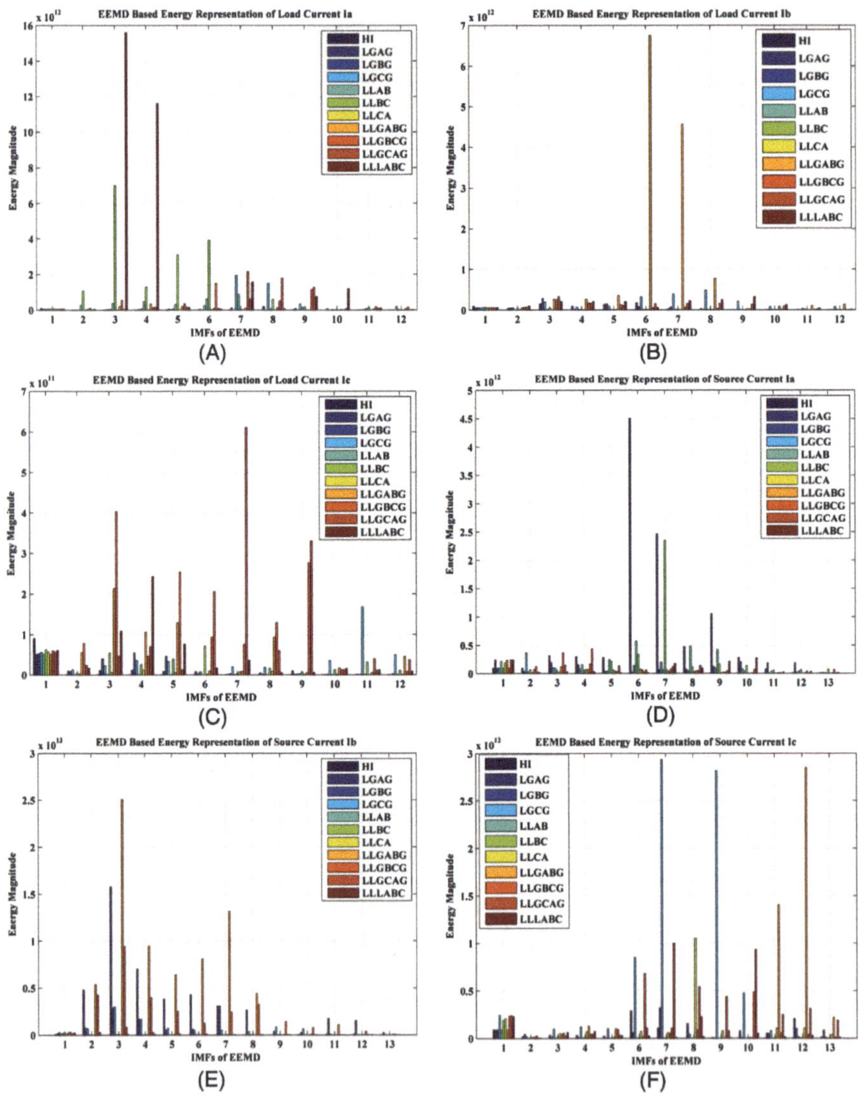

FIGURE 6.5 Energy distribution of IMFs of EEMD for (A) Load current Ia, (B) Load current Ib, (C) Load current Ic, (D) Source current Ia, (E) Source current Ib, (F) Source current Ic, (G) Load voltage Va, (H) Load voltage Vb, (I) Load voltage Vc, (J) Source voltage Va, (K) Source voltage Vb, and (L) Source voltage Vc.

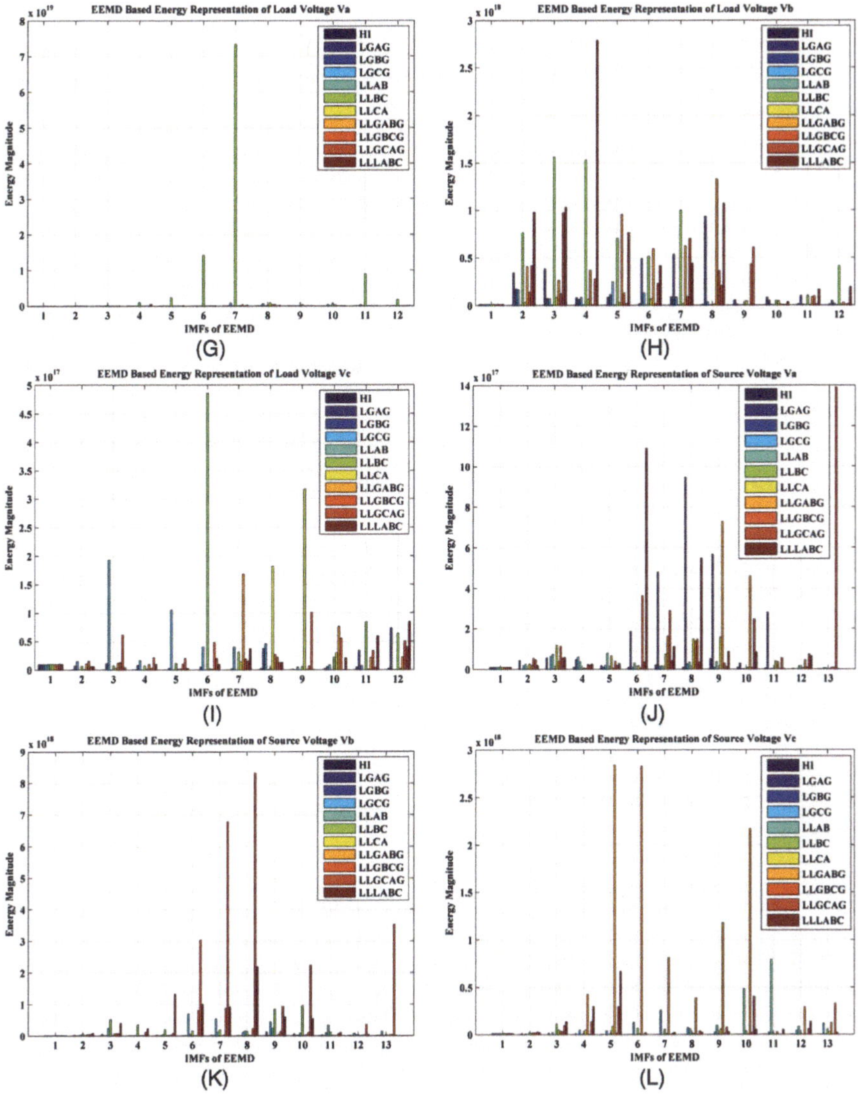

FIGURE 6.5 *(Cont.)*

Where, $\omega \in R^d$, $<\omega, S>$ = inner (dot) product of ω and S, and b is real.

The Eq. (6.1) shows the best hyperplane to segregate the dataset into two classes and further to evaluate ω and b that minimize $||\omega||$ such that for all training dataset (S_j, T_j):

$$T_j(<\omega, S> + b) \geq 1 \tag{6.2}$$

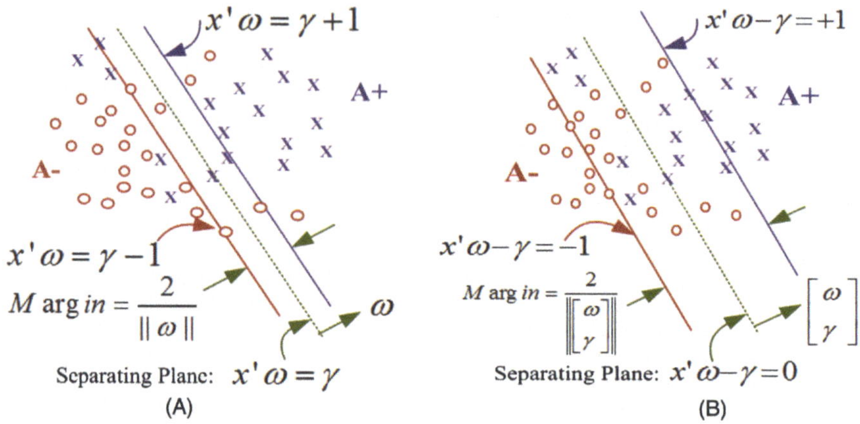

FIGURE 6.6 (A) The Standard SVM Classifier in the w-space of R^n: with some error margin $\frac{2}{\|w\|}$, and (B) The PSVM in the (ω, λ) space of R^{n+1} : with some error margin $\frac{2}{\left\|\begin{bmatrix}\omega\\ \gamma\end{bmatrix}\right\|}$.

Where, SVs are S_j on hyperplane boundary, where

$$T_j(<\omega, S> + b) = 1. \tag{6.3}$$

Therefore, the data-classification is categorized using optimum solution of ω and b as:

$$class(Pnew) = sign(<\omega, Pnew> + b) \tag{6.4}$$

Where, $Pnew$ = dataset used for classification.

2.5 Proximal support vector machine (PSVM)

PSVM mainly used for data classification [17–20]. It is an ameliorated edition of an available standard SVM. SVM is based on supervised learning theory where it is fed with data specimens (inputs) with associated labels (target value). These data specimen can be analyzed as per the breadth of hyperplane, which categorizes data specimen into two categories with a margin as shown in Fig. 6.6A, while in PSVM data specimen are classified based on its proximity as represented in Fig. 6.6B.

Fig. 6.7 represented the stepwise procedure of PSVM implementation. PSVM is taken as m data points in R^n represented using two matrices as matrix Q $(m \times n)$ and diagonal matrix G $(m \times m)$ with label of ±1. The label ±1 represents the class of each row of matrix Q. The mathematical implementation for linear classifier using PSVM is shown in following steps:

Step#1: Express M as:

$$M = G[Q - \varepsilon] \tag{6.5}$$

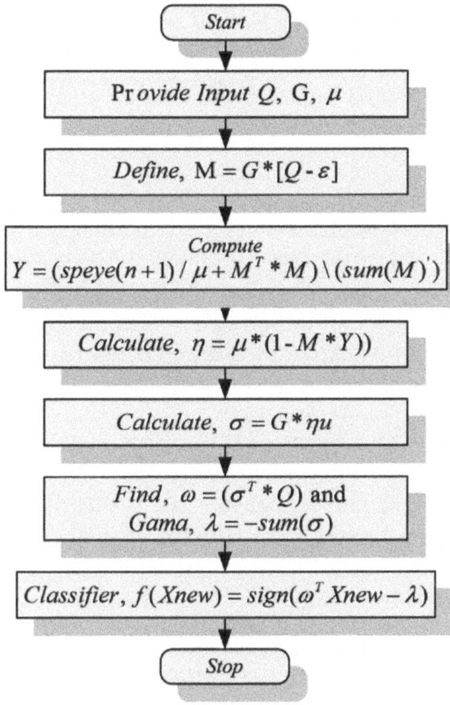

FIGURE 6.7 Flowchart for PSVM implementation [16].

Where $\varepsilon = m \times 1$ which is a vector of 1 and calculates η using positive different μ values:

$$\eta = \mu\left(\left(1 - M\left(\frac{1}{\mu} + M'M\right)^{-1}\right)M'\right)\varepsilon \tag{6.6}$$

or η can also be calculated as:

$$\eta = \mu * \left(1 - \left(M * \vartheta\right)\right) \tag{6.7}$$

Where ϑ is represented as: $\vartheta = \left(\frac{speye(n+1)}{\mu + M'*M}\right)\backslash Y$ and $Y = sum(M)'$.

The optimal value for μ is decided according to expert skills which varies from 10^6 to 0.01. It is suggested that bigger value of μ provides superior fitting for training data, mostly $\mu = 1$ functions best.

Step#2: Calculate (ω, λ)

$$\omega = Q' * G\eta; \text{ and } \lambda = -\varepsilon' * G\eta \text{ and } \tau = \eta/\mu \tag{6.8}$$

The maximization of the margin is achieved among the bounding planes in accordance to location (λ) and orientation (ω) with respect to origin.

Step#3: The new labeled or unlabeled dataset of *Xnew* is categorized as follow:

$$\left(Xnew'\omega - \lambda\right) = \begin{cases} > 0, & then \quad Xnew \in A+; \\ < 0, & then \quad Xnew \in A-; \\ = 0, & then \quad \text{Xnew} \in A + or \ \text{Xnew} \in A-; \end{cases} \tag{6.9}$$

After proper performance of training phase, testing is performed with unlabeled dataset. The identified class of the tested dataset is defined by Eq. (6.10) which is the function of $''\lambda''$ and $''\omega''$.

$$f(Xnew) = sign\left(\omega' * Xnew - \lambda\right) \tag{6.10}$$

Where, $f(Xnew)$ = negative, then dataset is classified as class#1 $(A-)$, and $f(x)$ = positive, then dataset is classified as class#2 $(A+)$.

For multiple number of classifications, multiple binary PSVM classifiers are used in a combination subsequently.

2.6 SVM and PSVM based transmission line fault classification model formation

Based on proposed approach as shown in Fig. 6.1 of Section 2.1, the TL fault detection and diagnosis model using SVM (TFDDS) and PSVM (TFDDP) is shown in Figs. 6.8 and 6.9, respectively.

Both TFDDS and TFDDP include 10-binary SVM and PSVM classifiers respectively which are utilized to classify the 11-condition, that is, healthy condition (HI) and the 10 transmission line faults (F1–F10) (F1:LGAG, F2:LGBG, F3:LGCG, F4:LLGABG, F5:LLGBCG, F6:LLCAG, F7:LLAB, F8:LLBC, F9:LLCA, and F10:LLLABC) scenario.

In the FDD procedure using 10-binary classifiers, SVM model#1/PSVM model#1 is trained using whole datasets of 11-cases to classify the HI condition form 10 faulty conditions (F1–F10). For indicating the HI-condition, output of the model#1 is represented by -1, otherwise 1. Now, faulty datasets (F1–F10) are utilized to train the binary SVM model#2/PSVM model#2 to categorize the ground faults (F1-F6) from the short circuited fault (F7–F10). For demonstrating the "ground faults" condition from "short circuited," the output of model#2 is represented by -1, otherwise 1. Now, short-circuited faulty data of F7–F10 are utilized to train the SVM modle#3/PSVM model#3 to categorize "F7–F9" faults from "F10" fault. For demonstrating the triple line short-circuited fault conditions (F10), the output of SVM model#3/PSVM model#3 is represented by -1; otherwise 1 for F7–F9 condition. Thereafter, double line short-circuited faulty dataset of F7–F9 faults are utilized to train the SVM model#4/PSVM model#4

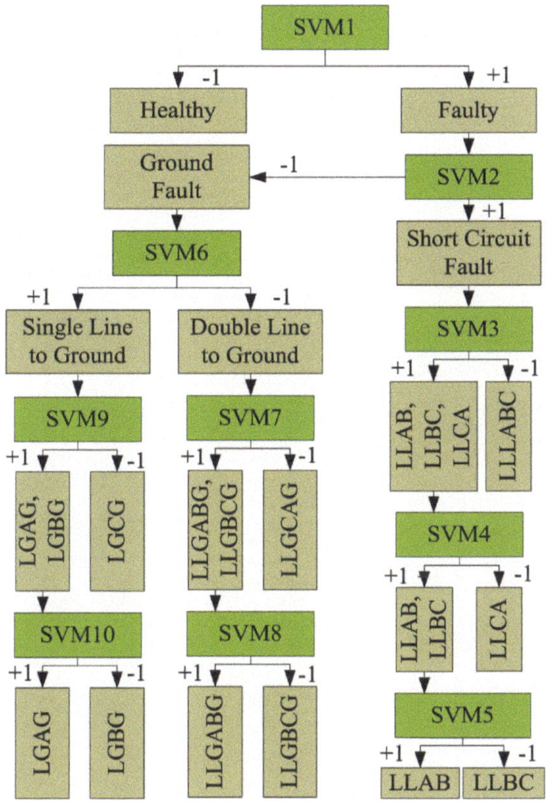

FIGURE 6.8 Classification approach through 10 SVM models.

to categorize the "F7 and F8" faults from "F9" fault. For demonstrating "F9" fault condition, the output of SVM model#4/PSVM model#4 is represented by -1; otherwise 1. Now, "F7 and F8" faults dataset are utilized to train the SVM model#5/PSVM model#5 to categorize "F7" fault condition from "F8" fault. For demonstrating "F8" fault condition, the output of SVM model#5/PSVM model#5 is represented by -1, otherwise 1. In this way, FDD of short-circuited line condition has been completed.

Now further classification of ground fault conditions, the ground fault dataset of (F1–F6) faults is utilized to train the SVM model#6/PSVM model#6 to categorize "F1–F3 (single line to ground)" faults from "F4–F6 (double line to ground)" faults. For demonstrating the "F4–F6" fault conditions, the output of SVM model#6/PSVM model#6 is represented by -1, otherwise 1. Thereafter, double-line-to-ground fault dataset of F4-F6 faults are utilized to train the SVM model#7/PSVM model#7 to categorize "F6" fault condition from "(F4 and F5)" faults condition. For demonstrating "F6" fault condition, the output of SVM

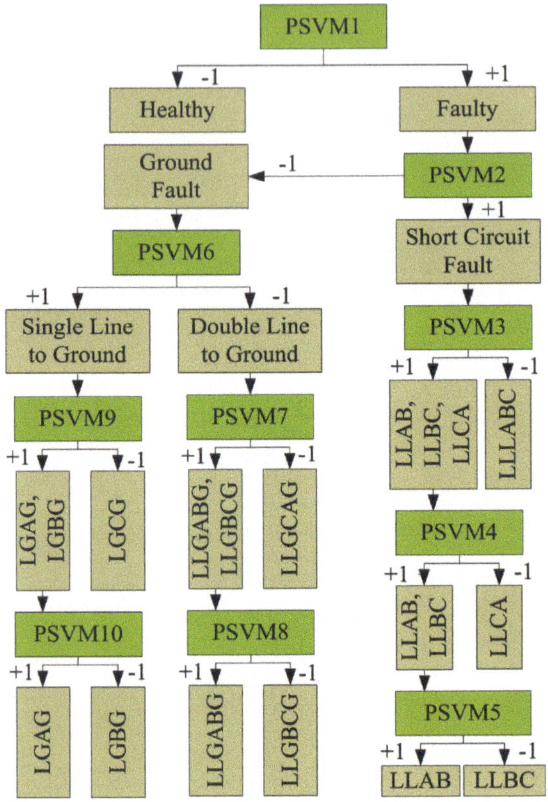

FIGURE 6.9 Classification approach through 10 PSVM models.

model#7/PSVM model#7 is represented by -1, otherwise 1. Now, the faulty dataset of F4 and F5 are utilized to train the SVM model#8/PSVM model#8 to categorize "F5" fault from "F4" fault condition. For demonstrating "F5" fault condition, the output of SVM model#8/PSVM model#8 is represented by -1, otherwise 1. Thereafter, single-line-to-ground fault dataset of F1-F3 are utilized to train the SVM model#9/PSVM model#9 to categorize "F1 & F2" fault form "F3" fault condition. For designating "F3" fault condition, the output of SVM model#9/PSVM model#9 is represented by -1, otherwise 1. Finally, F1 and F2 dataset are utilized to train the SVM model#10/PSVM model#10 to categorize "F1" fault from "F2" fault condition. For demonstrating the "F2" fault condition apart from "F1", the output of SVM model#10/PSVM model#10 is represented by -1, otherwise 1.

Hence, Table 6.2 represents the summary of output codification of all developed 10-SVM/PSVM models. The performance measures of each SVM as well as PSVM diagnostics model are analyzed with respect to the model parameter optimization.

TABLE 6.2 Summary of output codification of all developed 10-SVM/PSVM models.

Fault category	SVM/ PSVM model #1	SVM/ PSVM model #2	SVM/ PSVM model #3	SVM/ PSVM model #4	SVM PSVM model #5	SVM/ PSVM model #6	SVM/ PSVM model #7	SVM/ PSVM model #8	SVM/ PSVM model #9	SVM/ PSVM model #10
Healthy	−1	Θ	Θ	Θ	Θ	Θ	Θ	Θ	Θ	Θ
LLAB	+1	+1	+1	+1	+1	Θ	Θ	Θ	Θ	Θ
LLBC	+1	+1	+1	+1	−1	Θ	Θ	Θ	Θ	Θ
LLCA	+1	+1	+1	−1	Θ	Θ	Θ	Θ	Θ	Θ
LLLABC	+1	+1	−1	Θ	Θ	Θ	Θ	Θ	Θ	Θ
LGAG	+1	−1	Θ	Θ	Θ	+1	Θ	Θ	+1	+1
LGBG	+1	−1	Θ	Θ	Θ	+1	Θ	Θ	+1	−1
LGCG	+1	−1	Θ	Θ	Θ	+1	Θ	Θ	−1	Θ
LLGABG	+1	−1	Θ	Θ	Θ	−1	+1	+1	Θ	Θ
LLGBCG	+1	−1	Θ	Θ	Θ	−1	+1	−1	Θ	Θ
LLGCAG	+1	−1	Θ	Θ	Θ	−1	−1	Θ	Θ	Θ

Θ, not in used.

3 Results and discussions

3.1 SVM based transmission line fault classification

Based on the proposed approach (Fig. 6.1), the complete structure of the transmission line fault diagnosis model has been presented in Fig. 6.8 by using the SVM approach. The MATLAB based code has been utilized to design 80 diagnostic models of 8 different categories (i.e., category1: RCL-raw current load, category2: RCS-raw current source, category3: RVL-raw voltage load, category4: RVS-raw voltage source, category5: IMFs of raw current load, category6: IMFs of the raw current source, category7: IMFs of raw voltage load, and category8: IMFs of the raw voltage source data). The diagnosis accuracy of each model has been listed in Table 6.3. The overall diagnostic accuracy of SVM models under category#6 is 99.992% and 89.099% for the training and testing phase, respectively, which is higher than the developed models of other categories. The graphical results for all models are shown in Fig. 6.10. The detailed classification accuracy analysis (class-wise) for each SVM model of category#6 has been represented in Table 6.4.

3.2 PSVM based transmission line fault classification

Fig. 6.9 represents the step-wise procedure for fault diagnosis using the PSVM approach. The MATLAB based code for PSVM technique has been formulated to design and analyze 80 binary classifiers of 8 categories (i.e., category1: RCL-raw current load, category2: RCS-raw current source, category3: RVL-raw voltage load, category4: RVS-raw voltage source, category5: IMFs of raw current load, category6: IMFs of the raw current source, category7: IMFs of raw voltage load, and category8: IMFs of raw voltage source data) as shown in Table 6.5. After analyzing Table 6.5, it is found that the highest diagnostic accuracy for the transmission line is with category#6. The overall efficiency of category#6 based PSVM models has been found to be 97.349% (for training phase) and 93.017% (for testing phase), which is comparatively higher as compared with other PSVM models as mentioned in Table 6.5. Training was done for the various value of Nu (range of Nu = 6^{10} to 0.01) to check the transmission line fault diagnosis accuracy. In this case, fault identification is higher at Nu = 1, 0.1, & 0.01 as represented in Fig. 6.11A–J) for all 10 models. The detailed analysis of classification accuracy for category#6 is presented in Table 6.6, where calculated value of weight (w) and gama (G) for PSVM1 to PSVM10 with variation of Nu for each class has been represented individually (Fig. 6.12).

3.3 Comparative results analysis of SVM and PSVM based fault classification models

After implementation of SVM and PSVM approaches, 160 number of diagnostics models (80 diagnostic models by SVM and 80 diagnostic models by

TABLE 6.3 SVM based performance analysis.

Dataset used	SVM phase	SVM1	SVM2	SVM3	SVM4	SVM5	SVM6	SVM7	SVM8	SVM9	SVM10	Average
Category1: RCL	Train	90.909	65.533	75	66.667	50	50	66.667	50	66.667	50.083	63.153
	Test	90.909	41.025	75	66.667	49.75	46194	66.667	49.917	66.667	50	60.28
Category2: RCS	Train	90.909	65	75	66.667	50	50	66.667	50	66.667	50	63.091
	Test	90.909	40.917	75	66.667	49.958	46.861	66.667	49.625	66.667	50	60.327
Category3: RVL	Train	90.909	64.65	75	66.667	50	50.486	66.667	50.542	66.667	50.208	63.18
	Test	90.909	42.55	75	66.667	50	50	66.667	50	66.667	50	60.846
Category4: RVS	Train	90.909	64.6	75	66.667	50.125	50	66.667	51.375	66.667	55.125	63.713
	Test	90.909	42.258	75	66.667	50	41.069	66.667	50	66.667	50	59.924
Category5: EEMDCL	Train	100	100	100	100	100	100	100	100	100	100	100
	Test	87.599	57.95	58.333	88.889	83.333	51.181	66.667	54.25	64.722	44.125	65.705
Category6: EEMDCS	Train	99.917	100	100	100	100	100	100	100	100	100	**99.992**
	Test	96.644	84.65	87.361	85.208	86.556	86.556	89.472	95.958	81.111	98.917	**89.099**
Category7: EEMDVL	Train	99.061	99.252	100	100	100	100	100	100	99.667	100	99.798
	Test	94.561	76.492	83.583	81.389	90.292	64.319	72.417	73.292	75.111	70.042	78.15
Category8: EEMDVS	Train	99.712	99.442	100	100	100	98.931	100	100	100	100	99.808
	Test	97.22	86.042	90.917	80.806	94.875	76.194	76.694	71.833	87.806	88.833	85.122

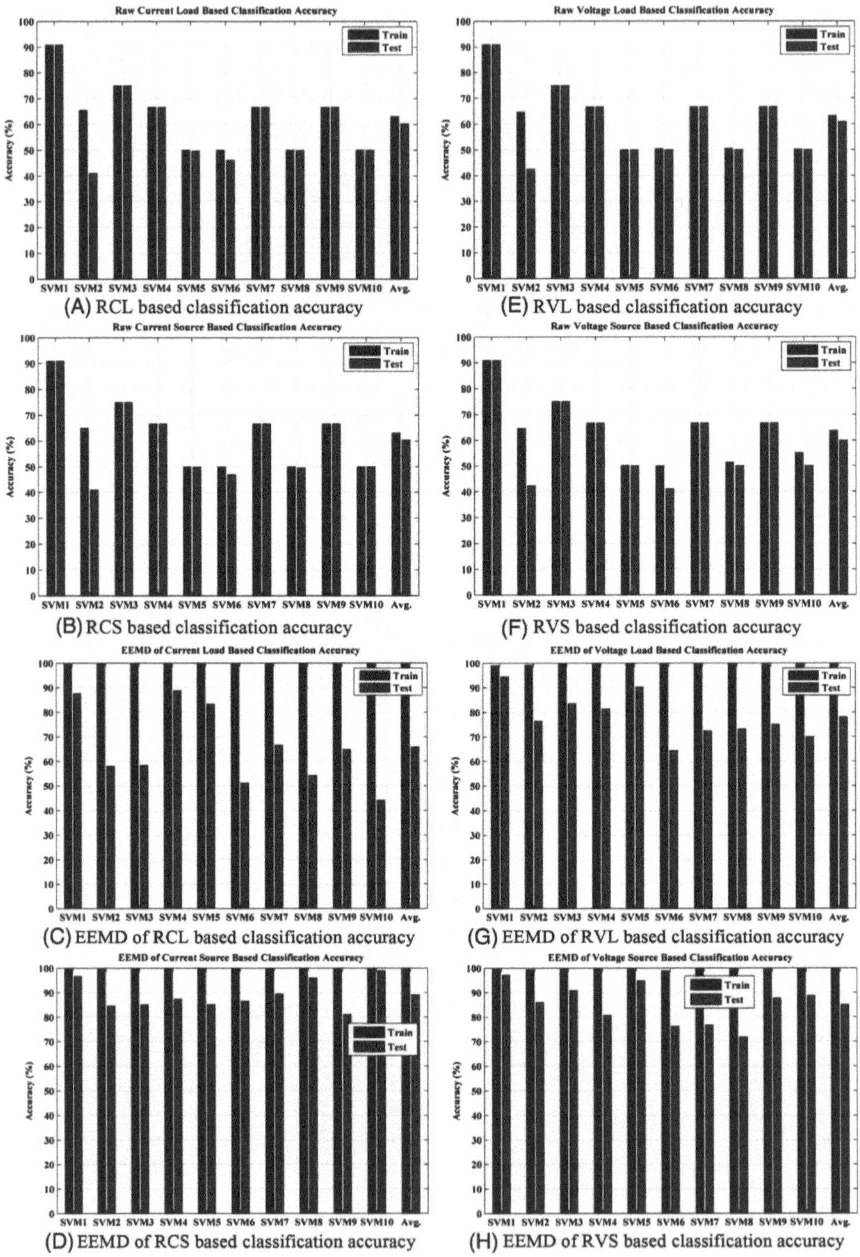

(A) RCL based classification accuracy

(E) RVL based classification accuracy

(B) RCS based classification accuracy

(F) RVS based classification accuracy

(C) EEMD of RCL based classification accuracy

(G) EEMD of RVL based classification accuracy

(D) EEMD of RCS based classification accuracy

(H) EEMD of RVS based classification accuracy

FIGURE 6.10 **SVM based transmission line fault diagnosis accuracy analysis.**

TABLE 6.4 Class-wise classification accuracy analysis for SVM1–SVM10 based on "EEMDCS."

SVM model/ data sets used	Transmission line operating condition	MSE	RMSE	Diagnosis accuracy in training phase (%)		Diagnosis accuracy in testing phase (%)	
SVM model1	Healthy condition	0.0008	0.02881	1200/1200	99.917	1189/1200	96.644
	Faulty condition			11989/12000		11568/12000	
SVM model2	Ground Fault	0.0	0.0	4800/4800	100	4163/4800	84.65
	Short Circuit Fault			7200/7200		6842/7200	
SVM model3	LLAB,LLBC,LLCA	0.0	0.0	3600/3600	100	3263/3600	85.104
	LLLABC			1200/1200		822/1200	
SVM model4	LLAB,LLBC	0.0	0.0	2400/2400	100	2196/2400	87.361
	LLCA			1200/1200		949/1200	
SVM model5	LLAB	0.0	0.0	1200/1200	100	923/1200	85.208
	LLBC			1200/1200		1122/1200	
SVM model6	Double line to ground	0.0	0.0	3600/3600	100	3316/3600	86.556
	Single line to ground			3600/3600		2916/3600	
SVM model7	LLGABG,LLGBCG	0.0	0.0	2400/2400	100	2197/2400	89.472
	LLGCAG			1200/1200		1024/1200	
SVM model8	LLGABG	0.0	0.0	1200/1200	100	1178/1200	95.958
	LLGBCG			1200/1200		1125/1200	
SVM model9	LGAG,LGBG	0.0	0.0	2400/2400	100	1996/2400	81.111
	LGCG			1200/1200		924/1200	
SVM model10	LGAG	0.0	0.0	1200/1200	100	1198/1200	98.917
	LGBG			1200/1200		1176/1200	
			Average accuracy (%)		**99.992**		**89.099**

TABLE 6.5 PSVM based performance analysis of recorded raw data and its EEMD.

Dataset used	PSVM Phase	PSVM1	PSVM2	PSVM3	PSVM4	PSVM5	PSVM6	PSVM7	PSVM8	PSVM9	PSVM10	Average
Category1: RCL	Train	82.434	68.217	75	66.667	50	54.375	66.667	50	66.667	50	63.003
	Test	74.079	60.867	75	66.667	49.875	50.639	66.667	49.208	66.667	49.792	60.946
Category2: RCS	Train	89.948	64.167	75	66.667	59.875	50.847	66.667	50	66.667	50	63.984
	Test	86.047	60.817	75	66.667	49.875	49.931	66.667	49.875	66.667	49.791	62.134
Category3: RVL	Train	81.647	60.009	75	70.75	50	50	69.111	50	66.667	50	62.318
	Test	75.964	60.008	75	69.083	49.958	49.958	67.278	49.917	66.667	49.958	61.379
Category4: RVS	Train	87.092	65.783	62.792	72.083	50	50.097	66.66	52.935	66.66	50	62.410
	Test	82.074	62.617	60.313	70.333	49.958	49.955	66.66	50.958	66.66	49.792	60.932
Category5: EEMDCL	Train	90	72.975	100	100	100	71.986	93.333	100	91.111	100	91.941
	Test	84.848	56.425	83.333	84.611	83.333	61.111	87.778	93.543	89.083	83.333	80.739
Category6: EEMDCS	Train	90.909	91.158	97.146	99.556	100	94.722	100	100	100	100	**97.349**
	Test	90.909	89.292	92	94.194	92.167	93.986	95.778	90.958	90.889	100	**93.017**
Category7: EEMDVL	Train	87.891	68.475	86.854	96.389	87.583	70.153	87.056	90.292	95.417	76.083	84.619
	Test	87.284	66.517	82.375	91.194	86.292	65.778	82.25	89.792	91.876	74.042	81.739
Category8: EEMDVS	Train	89.054	66.525	99.604	96.305	80.666	83.403	94.694	95.5	100	100	90.575
	Test	88.054	63.925	94.625	93.25	78.625	79.597	89.194	89.25	98.194	96.417	87.113

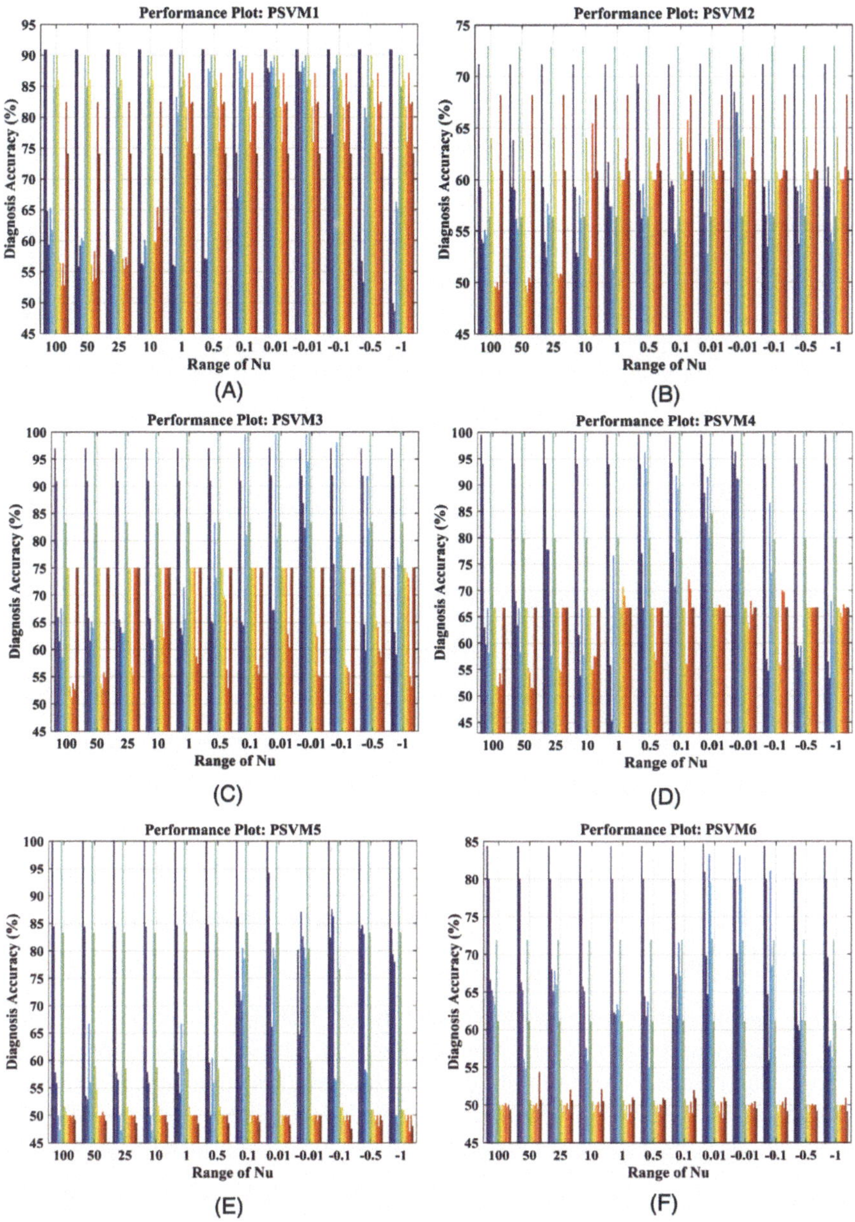

FIGURE 6.11 PSVM based transmission line fault diagnosis accuracy analysis with variation of Nu value for (A) PSVM1, (B) PSVM2, (C) PSVM3, (D) PSVM4, (E) PSVM5, (F) PSVM6, (G) PSVM7, (H) PSVM8, (I) PSVM9, and (J) PSVM10.

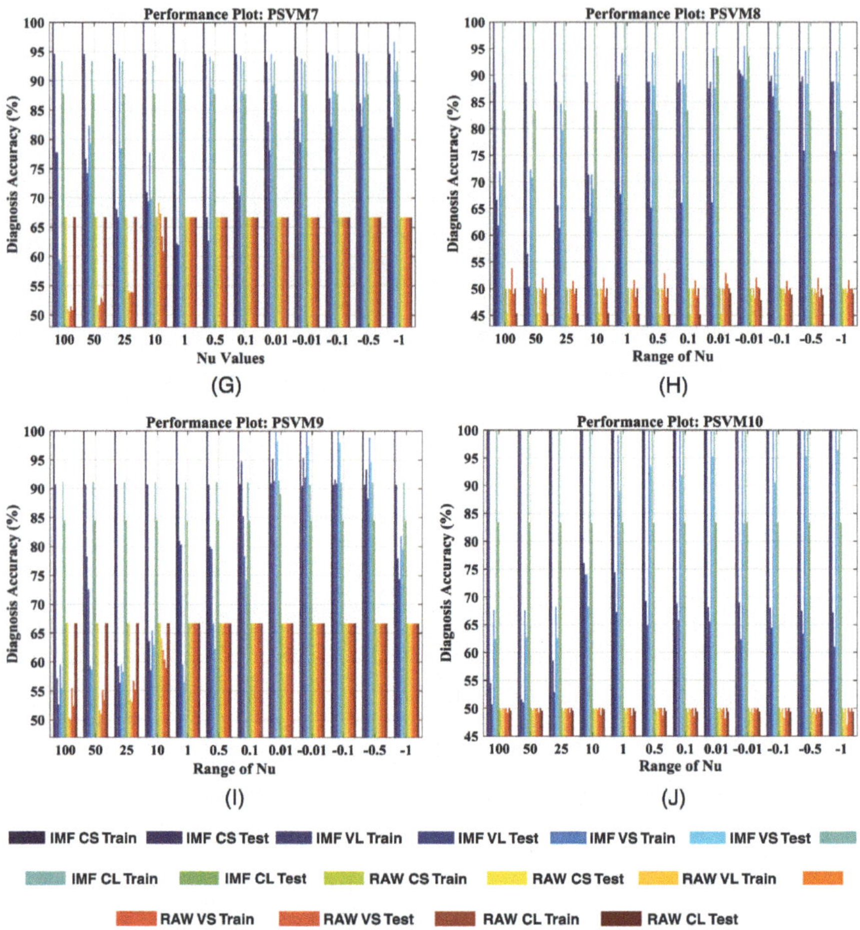

FIGURE 6.11 (*Cont.*)

PSVM) have been implemented by using different input variables of 8 different categories and then classification accuracy have been evaluated for analyzing the classification compatibility of both classification approaches (SVM and PSVM). After detailed comparison, it is concluded that only one specific category will provide best results that is known as category#6 based PSVM models as shown in Fig. 6.13. PSVM based transmission line fault diagnosis model has acceptable diagnostic accuracy than SVM based models.

4 Conclusion

The presented chapter shows the advantages of the EEMD data decomposition technique in transmission line fault diagnosis by using PSVM and SVM based classifiers. A total of 160 numbers of intelligent classifiers have been designed

TABLE 6.6 Class-wise classification accuracy analysis for PSVM1–PSVM10 based on "EEMDCS."

PSVM model/ data sets used	Type of transmission line operating condition	MSE	RMSE	Diagnosis accuracy in training phase (%)		Diagnosis accuracy in testing phase (%)	
PSVM model1	Healthy condition	0.09091	0.3015	1099/1200	90.909	1080/1200	90.909
	Faulty condition			10901/12000		10920/12000	
PSVM model2	Ground fault	0.0884	0.2974	4475/4800	91.158	4298/4800	89.292
	Short circuit fault			6464/7200		6417/7200	
PSVM model3	LLAB,LLBC,LLCA	0.02854	0.1689	3517/3600	97.146	3382/3600	92
	LLLABC			1146/1200		1034/1200	
PSVM model4	LLAB,LLBC	0.00444	0.0666	2399/2400	99.556	2280/2400	94.194
	LLCA			1185/1200		1111/1200	
SVM model5	LLAB	0.0	0.0	1200/1200	100	1039/1200	92.167
	LLBC			1200/1200		1173/1200	
PSVM model6	Double line to ground	0.05278	0.2297	3505/3600	94.722	3394/3600	93.986
	Single line to ground			3315/3600		3373/3600	
PSVM model7	LLGABG,LLGBCG	0.0	0.0	2400/2400	100	2304/2400	95.778
	LLGCAG			1200/1200		1144/1200	
PSVM model8	LLGABG	0.0	0.0	1200/1200	100	1082/1200	90.958
	LLGBCG			1200/1200		1101/1200	
PSVM model9	LGAG,LGBG	0.0	0.0	2400/2400	100	2198/2400	90.889
	LGCG			1200/1200		1074/1200	
PSVM model10	LGAG	0.0	0.0	1200/1200	100	1200/1200	100
	LGBG			1200/1200		1200/1200	
			Average accuracy (%)		**97.349**		**93.017**

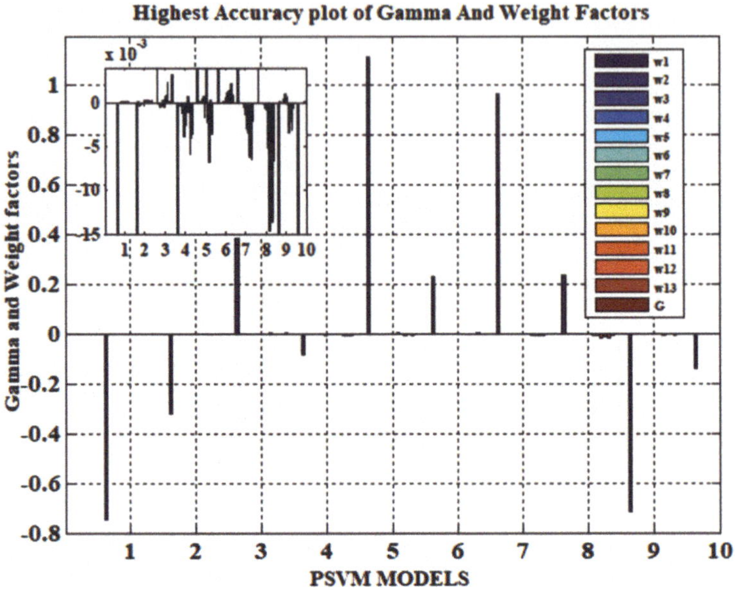

FIGURE 6.12 Calculated weight (w) and gamma (G) value for PSVM1–PSVM10 with variation of Nu.

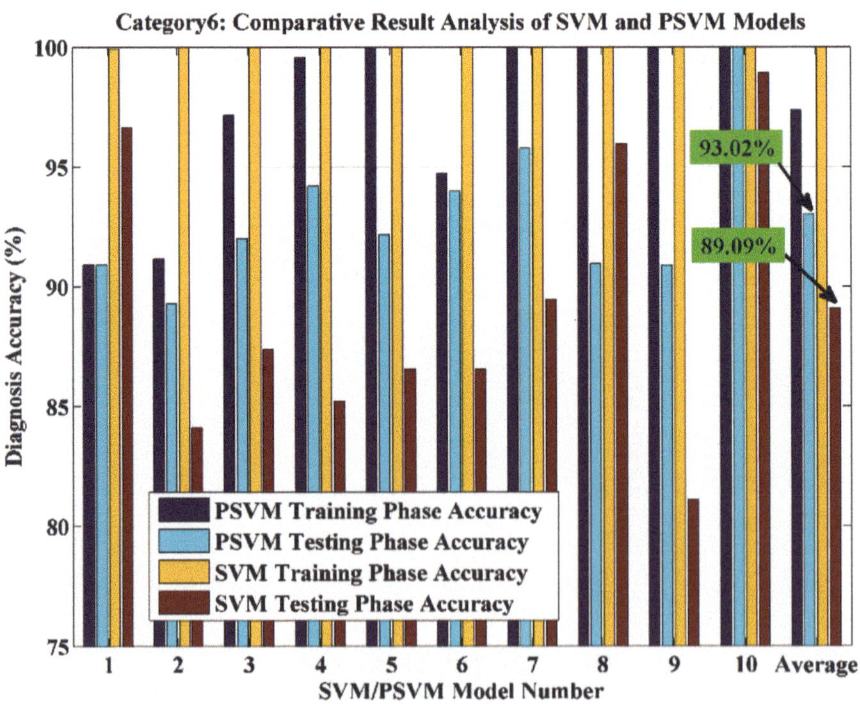

FIGURE 6.13 Comparative result analysis of SVM and PSVM.

and implemented in this study, out of which, the 80 classifiers are based on the raw dataset and 80 classifiers are based on the EEMD technique. PSVM based obtained results of eight different categories (category1–category8) are compared with SVM based results which show the superiority of the proposed PSVM classifiers. In term of processing time, PSVM is also faster than SVM and tuning of model parameters are easy. Furthermore, both SVM and PSVM models are developed, validated, and tested for classification of the faults occurring in the power transmission line. The results suggest that PSVM proves to be effective and takes much shorter training/testing time as compared with SVM.

For the future scope, the proposed approach can be adopted for online, non-intrusive FDD of smart-grid applications, which will reduce the O&M cost and increase the operating life span of the equipment as well as the system.

References

[1] Y.Q. Chen, O. Fink, G. Sansavini, Combined fault location and classification for power transmission lines fault diagnosis with integrated feature extraction, IEEE Trans. Ind. Electron. 65 (1) (2018) 561–569, doi: 10.1109/TIE.2017.2721922.

[2] H. Malik, S. Mishra, Artificial neural network and empirical mode decomposition based imbalance fault diagnosis of wind turbine using TurbSim, FAST and Simulink, IET Renew. Power Generat. 11 (6) (2017) 889–902, doi: 10.1049/iet-rpg.2015.0382.

[3] H. Malik, S. Mishra, Application of GEP to investigate the imbalance faults in direct-drive wind turbine using generator current signals, IET Renew. Power Generat. 11 (6) (2017) 889–902, doi: 10.1049/iet-rpg.2016.0689.

[4] N. Fatema, H. Malik, A. Iqbal, Data driven intelligent model for sales prices prediction and monitoring of a building, Soft Computing in Condition Monitoring and Diagnostics of Electrical and Mechanical Systems, Springer, 2019, pp. 407–421, doi: https://doi.org/10.1007/978-981-15-1532-3_18.

[5] H. Malik, S. Mishra, Application of GEP to investigate the imbalance faults in direct-drive wind turbine using generator current signals, IET Renew. Power Generat. 12 (3) (2018) 279–291, doi: 10.1049/iet-rpg.2016.0689.

[6] A.K. Yadav, H. Malik, S.S. Chandel, Selection of most relevant input parameters using WEKA for artificial neural network based solar radiation prediction models, Renew. Sustain. Energy Rev. 31 (2014) 509–519, doi: https://doi.org/10.1016/j.rser.2013.12.008.

[7] R.K. Aggarwal, Y. Aslan, A.T. Johns, New concept in fault location for overhead distribution systems using superimposed components, IET Gener. Transm. Distrib. 144 (3) (1997) 309–316.

[8] C.K. Jung, J.B. Lee, X.H. Wang, Y.H. Song, Wavelet based noise cancellation technique for fault location on underground power cables, Electric Power Syst. Res. 77 (10) (2007) 1349–1362.

[9] I. Sadinezhad, M. Joorabian, A new adaptive hybrid neural network and fuzzy logic based fault classification approach for transmission lines protection, in: IEEE Second International Power and Energy Conference, Johor Bahru, Malaysia, 2008, pp. 895–900.

[10] N.R. Babu, B.J. Mohan, Fault classification in power systems using EMD and SVM, Ain Shams Eng. J. 8 (2) (2017) 103–111, doi: https://doi.org/10.1016/j.asej.2015.08.005.

[11] O.A.S. Youssef, Combined fuzzy-logic wavelet-based fault classification technique for power system relaying, IEEE Trans. Power Deliv. 19 (2) (2004) 582–589, doi: 10.1109/TPWRD.2004.826386.

[12] H. Malik, R. Sharma, Transmission line fault classification using modified fuzzy Q learning, IET Generat. Transm. Distrib. 11 (16) (2017) 4041–4050, doi: 10.1049/iet-gtd.2017.0331.

[13] H. Malik, R. Sharma, EMD and ANN based intelligent fault diagnosis model for transmission line, J. Intell. Fuzzy Syst. 32 (4) (2017) 3043–3050, doi: 10.3233/JIFS-169247.

[14] A. Aggarwal, H. Malik, R. Sharma, Selection of most relevant input parameters using WEKA for artificial neural network based transmission line fault diagnosis model, in: Proceedings of the International Conference on Nanotechnology for Better Living, vol. 3, No. 1, 2016, pp. 176, doi: 10.3850/978-981-09-7519-7nbl16-rps-176

[15] H. Malik, A. Aggarwal, R. Sharma, Feature extraction using EMD and classification through probabilistic neural network for fault diagnosis of transmission line, in: Proceedings of IEEE First International Conference on Power Electronics, Intelligent Control and Energy Systems, Delhi, India, 2016, pp. 1–6, doi: 10.1109/ICPEICES. 2016.7853709

[16] A. Swetapadma, A. Yadav, A novel decision tree regression-based fault distance estimation scheme for transmission lines, IEEE Trans. Power Deliv. 32 (1) (2017) 234–245.

[17] G. Fung, O.L. Mangasarian, Proximal support vector machine classifiers, in: Proceedings of KDD-2001: Knowledge Discovery and Data Mining, San Francisco, CA, 2001, pp. 77–86, doi: ftp://ftp.cs.wisc.edu/pub/dmi/tech-reports/01-02.pdf

[18] H. Malik, S. Mishra, Fault identification of power transformers using proximal support vector machine (PSVM), in: Proceedings of IEEE International Conference on Power Electronics (IICPE 2014), NIT Kurukshetra, India, 2014, doi: 10.1109/IICPE. 2014.7115842

[19] A.P. Mittal, H. Malik, V. Talur, S. Rastogi, External fault identification experienced by 3-phase induction motor using PSVM, in: Proceedings of IEEE International Conference on Power India (PIICON 2014), New Delhi, India, 2014, doi: 10.1109/POWERI. 2014.7117762

[20] H. Malik, S. Mishra, Proximal support vector machine (PSVM) based imbalance fault diagnosis of wind turbine using generator current signals,Energy Procedia, 90 (2016) 593–603, doi: https://doi.org/10.1016/j.egypro.2016.11.228

[21] Wu Zhaohua, Norden E. Huang, Ensemble empirical mode decomposition: A noise-assisted data analysis method, Adv. Adapt. Data Anal. 01 (01) (2011) 904–995, https://doi.org/10.1142/S1793536909000047.

Part B

Intelligent Data Analytics for Forecasting in Smart Grid

Chapter 7

Intelligent Data Analytics for Global Solar Radiation Forecasting for Solar Power Production Using Deep Learning Neural Network (DLNN)

1 Introduction

The energy potential such as solar energy (SE) plays a great role to reduce the adverse environmental effects including CO_2 level reduction. Due to the development and enhancement of recent technologies and reducing the cost, the SE will surely provide the future energy prospective. Approximate 60% of generated power is projected from renewable energy sources in 2040, which includes mainly solar and wind power potential apart of hydropower, biomass, geothermal, tidal, wave, hydrogen fuel cell, peats, etc. And the share of the total annual addition of the renewable energy in to the global power energy is growing day-by-day which is around 65% contribution of the total global power addition in 2018. The generation of solar energy is increasing day-by-day as in 1965 generated solar power was zero but now 0.39, 1.13, 33.68, 453.52, and 584.63 TWh in the year of 1990, 2000, 2010, 2017, and 2018, respectively [1].

Geographically, India is located between equator and tropic of cancer with different climatic zones and has huge solar potential [2]. Solar energy is converted into power through photovoltaic (PV) systems. The PV power varies from site to site due to varying solar radiation, which requires techno economic feasibility analysis of PV system in various climatic zones in India. In this aspect, Soni and Gakkhar [3] presented feasibility for PV technologies implementation in India and found that economic parameters are important for PV installation. The government launched Jawaharlal Nehru National Solar Mission (JNNSM) in 2010 with a target of 20,000 MW which has been revised to 175 GW by 2022. JNNSM is being implemented in three phases. The time line for finalizing phase 1, phase 2, and phase 3 is 2013, 2017, and 2022, respectively. The PV installed capacity plants of different states during phase 1 (2010–13) till January 31, 2014 is 2208.3655 [4,5] in which Rajasthan and Gujarat states have maximum installed PV system.

Intelligent Data-Analytics for Condition Monitoring. http://dx.doi.org/10.1016/B978-0-323-85510-5.00007-7

The power generated from the solar depends upon various climatic variables, which makes solar power time varying and intermittent. This imposes limitation on the PV generation capacity. Moreover, measurement of all variables itself is not economical procedure as well as data storage is also a burden on the system. So, forecasting of SR for the solar power evaluation is an alternative approach in this domain. Moreover, the optimum sizing and the installation of the solar photovoltaic system has made solar radiation as an important parameter in solar energy research. Solar radiation measuring equipments are costly, so they are not available at most of the meteorological stations. Also, missing records in the dataset have been found due to less accuracy of the available instruments. So, there is an urge of an accurate forecasting of solar radiation for such locations.

The PV power generation faces several problems in India such as lack of measured SR at different sites and optimum sizing of PV [6]. Estimation of SR on tilted surface is important for design process. The tilt angle affect dust deposition on PV [7] and incident SR on PV surface is maximized by tilt angle and tracking system [8]. The incident SR on horizontal surface for Indian cities is given in literature of Ref. [9]. It was shown that performance of tracking is highly dependent on latitude and thus varies with latitude. Moreover, this is not cost-effective and not much reliable to harness the information of the solar power. Therefore, SR forecasting comes in picture. In this regard, several SR forecasting models have been developed such as [9], [10–13], but most of models need further enhancement in term of performance and accuracy etc.

The objective of this study is to develop a cost-effective and highly accurate model to harness the information of the solar radiation for installation of a PV system and/or for planning the short term and long term goal for the power generation. In this study, deep neural network (DNN) based short term forecasting model is developed for the SR.

The chapter is organized into seven sections, which includes the introduction in Section 1 and data analysis for SRFP in Section 2. Section 3 represents the solar irradiance forecasting methods and Section 4 represents the study area and dataset collection used for study. The structure of the proposed model is shown in Section 5, which includes the DNN and performance evaluation measures for the developed model. The demonstrated results are shown in Section 6 and finally conclusion is shown in Section 7.

2 Data analysis for solar radiation forecasting and prediction (SRFP)

In this section, a data analysis related to recent invention has been performed by using several database platforms for solar radiation forecasting and solar radiation prediction. All analysis in this chapter are based on available open access database, although this is not sufficient for a crisp conclusion and decision making, but it can help the researcher to figure out and fix a target for further research at advance level in the SRFP area. In this analysis, several valuable

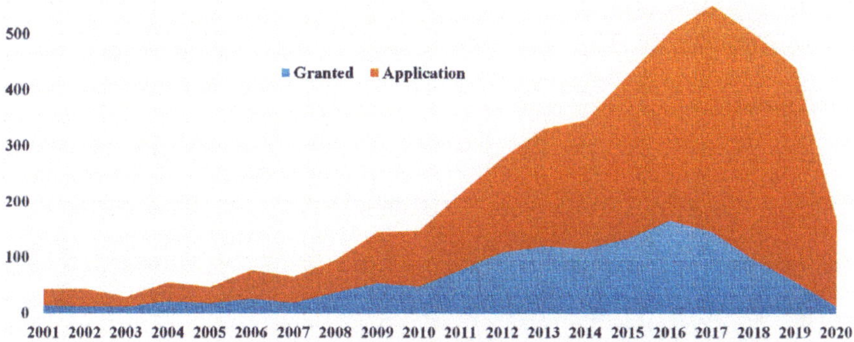

FIGURE 7.1 **New applications for inventions and number of granted applications.**

information such as profiles of innovation, innovations related to geographic area, available key technologies and methods, analysis of ownership, analysis of involved researchers or inventors and its co-inventors, significant inventions, analysis of market and its importance, innovative word related to the inventions, and analysis of each technology related to the solar radiation forecasting and prediction have been represented. For the data analytics of the recent research domain of SRFP, the focused innovative works are analyzed for the period of 2 decades (from 2001 to 2019 and some period of 2020), which shows the applied annual application versus granted application as shown in Fig. 7.1.

Based on the available information, it is concluded that most of work are available into six categories (i.e., inactive, active, PCT designated stage, PCT designated stage expired, undetermined, and pending applications). Generally, these innovative works belong to three main types, that is, invention type, design type, and utility type. Here maximum is invention type, that is, 94% plus of total inventions. Apart from this, available innovative work of STFP belongs to 11 main origins (i.e., China, United States, Japan, Korea, Germany, Russia, France, EPO, WIPO, India, and others). The participation from India is very negligible with respect to other territories. Moreover, as per the application trend (as shown in Fig. 7.2), China is at position one followed by United States and the trend is China-USA-Japan-WIPO-Korea-EPO-Australia-Russia-Germany-Canada. Moreover, the origin and protection of the technique and invention in the best five territories are EP (European patent office), CN (China), JP (Japan), KR (Korea), and US (United States). And according to the number and records form the best five territories, China is at the rank-1 position in SRFP area. Within the Chinese provinces, 10 sub-reasons have predominant research outcome which are: Beijing-Jiangsu-Guangdong-Shanghai-Shandong-Zhejiang-Tianjiang-Shaanxi-Sichuan-Anhui, as shown in Fig. 7.3.

In the area of SRFP, top 10 key technologies play an important role for the inventions. Broadly, these key technologies are: (1) KT1: systems or methods, (2) KT2: management, (3) KT3: circuit arrangements, (4) KT4: meteorology, (5) KT5: data processing or methods, (6) KT6: biological models, (7) KT7:

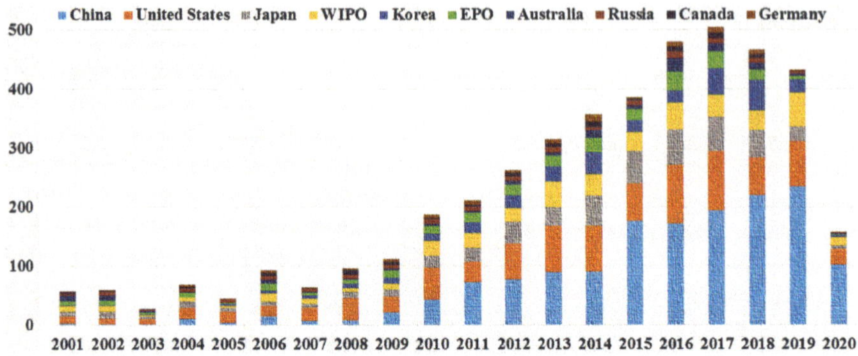

FIGURE 7.2 Invention trend in topmost countries in the area SRFP.

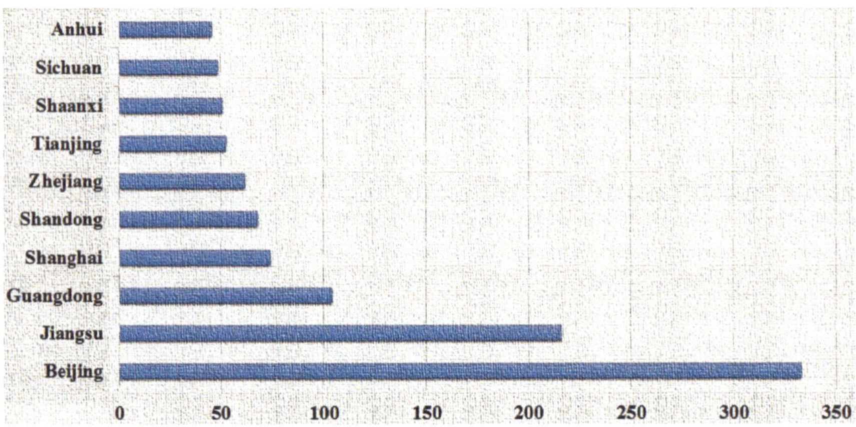

FIGURE 7.3 Top 10 Chinese provinces for SRFP inventions.

control & safety, (8) KT8: forecasting purposes, (9) KT9: network condition, and (10) KT10: battery or its load. The geographical distributions of these KT are shown in Fig. 7.4. According to the data analytics of KT, the top assignees for SRFP inventions are as follow: State Grid Corporation of China, Hitachi LTD, China Electric Power Research Institute, KK Toshiba, Chugoku Electric Power Company, Mitsubishi Electric Corp, Hohai University, North China Electricity Power University, Tianjin University, and Sekisui Chemical (as shown in Fig. 7.5). Moreover, the geographic representations of top assignees are shown in Fig. 7.6, which shows that most of key inventions are from only China.

The key take away from the data analytics of available inventions in the area of SRFP is that the market value of the invention is very high and key territory for the invention and implementation is China which plays a role of leader in this domain. This type of data analytics is very important for future planning and for mapping the ongoing research with respect to the industry players.

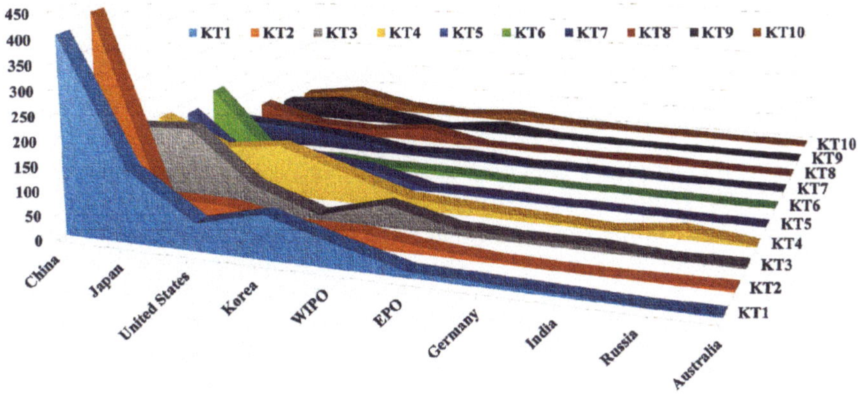

FIGURE 7.4 Geographic distribution of KTs.

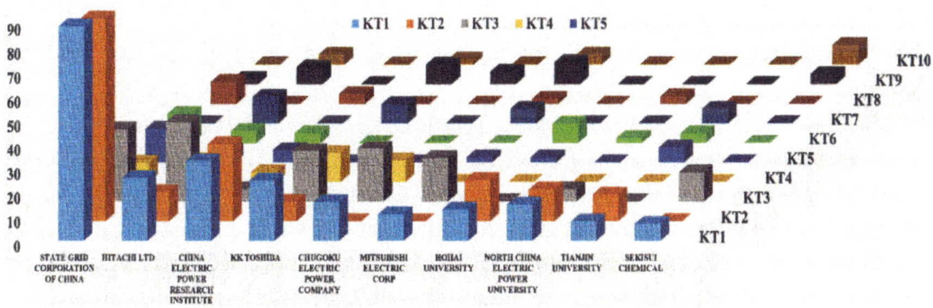

FIGURE 7.5 Top 10 assignee of SRFP.

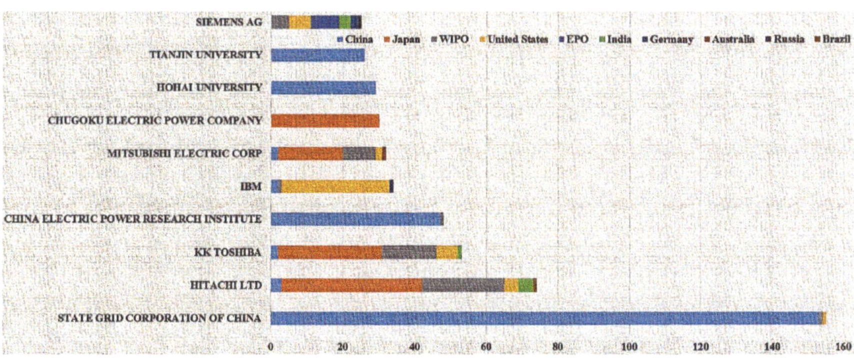

FIGURE 7.6 The geographic representation of top companies.

3 Solar irradiance forecasting methods

Solar forecasting techniques are of mainly three types: (1) numerical weather prediction (NWP) methods; (2) image based methods; and (3) statistical and machine learning (ML) methods. The third type methods are applicable for wide range of temporal horizons therefore becoming quite popular for forecasting.

4 Study area and dataset collection used for study

The Port Blair (latitude [°N] 11.61 and longitude [°E] 92.72 and altitude[m] 65) site in the south Andaman Island of the Andaman and Nicobar Islands, which is located in the Bay of Bengal, was considered for the study in this chapter as shown in the Fig. 7.7. Port Blair is covered by the sea and it has huge potential of solar potential throughout the year. Here daily average sunshine hours and daylight hours varies from 2.9 to 9.4 h and 11.5 to 12.8 h, respectively, throughout the year. Moreover, it has tropical climate with significant rainfall throughout the year. The average temperature, humidity, and pressure are 27.21°C, 82.05%, and 1003.52 hPa. According to historical record, the maximum global horizontal solar irradiance (GHI), direct normal irradiance (DNI), and diffuse horizontal irradiance (DHI) varies from 1085 to 1404, 729 to 909, and 560 to 895 (W/m^2), respectively, throughout the year of 2015 as shown in Fig. 7.8.

The historical data from meteorological department of India has been collected for the academic use. The eight main variables (i.e., irradiance, wind speed, wind direction, temperature, relative humidity, pressure, precipitation, and dew point) were collected at 1 min interval of time. The recorded data from January 1st, 2014 to December 31st, 2017 were used for the study. The further statistical analysis of the recorded data set are shown in Tables 7.1–7.4, which are minimum value, maximum value, mean, standard deviation (STD), and variance (VAR) (Figs. 7.9–7.12).

The historical data of SRFP is recorded per minute basis and assumed time duration for the recording of a day is [00:00:00] to [23:59:00] for each day. Generally, there are some missing values and spikes in the historical time series data due to unwanted weather condition and/or instrumental/operational/technical errors. Therefore, data preprocessing technique is implemented and purifies the historical data with filling the appropriate missing values and removing the spikes (if any). The purified per month historical data is shown in Fig. 7.13 for better understanding. In Fig. 7.13A–L, the historical and filled missing value data are represented for each month starting from January to December. The original historical data is represented by dark black color and filled missing value data is shown by gray color for better understanding, and visualization of the missing value pattern obtained by intelligent data analytics.

For further analysis of SRFP using DNN, per minute dataset has been used for training and testing phase. The historical data is recorded in 1-min interval. Therefore, total data samples for 1 month are around 44,640 and around 5,35,680 in a year, which are divided into two sub-dataset (training data~70% and testing data~30%).

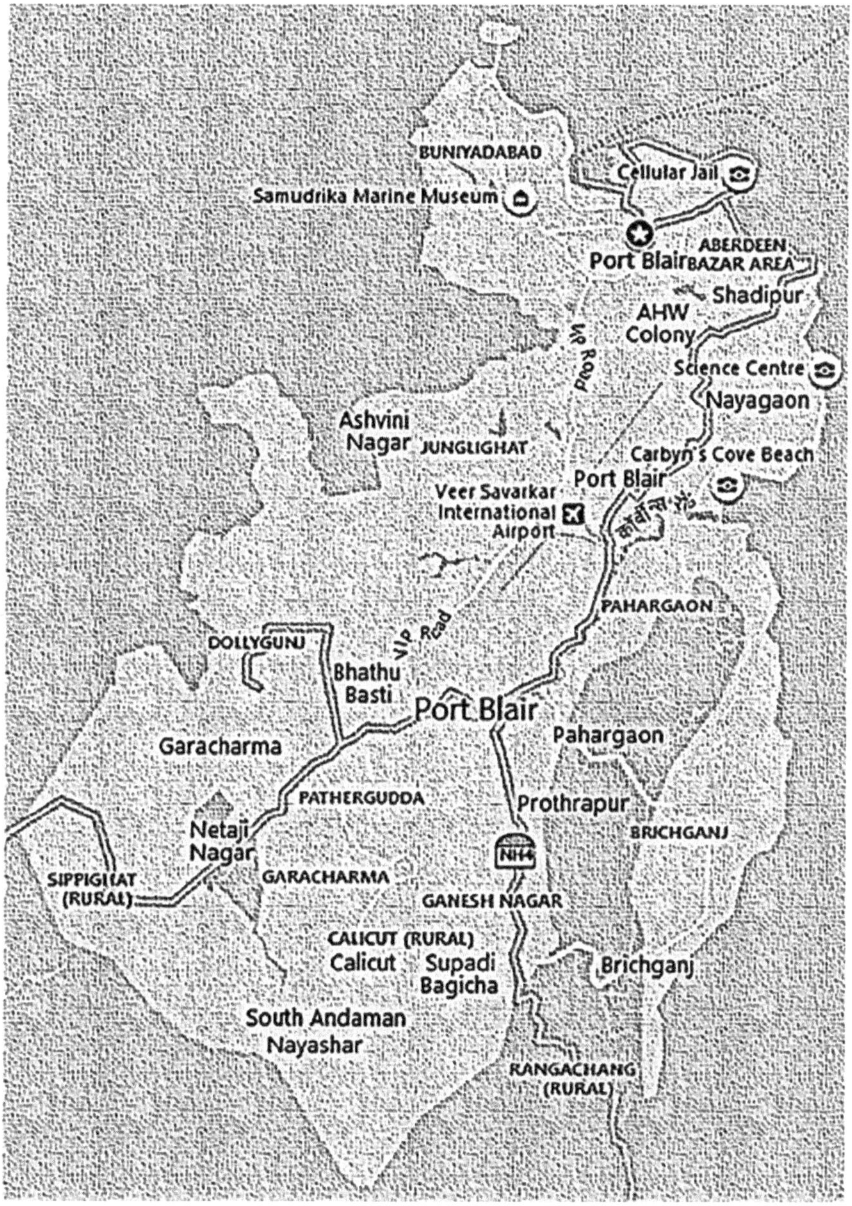

FIGURE 7.7 Location of Port Blair (in *dark white color*).

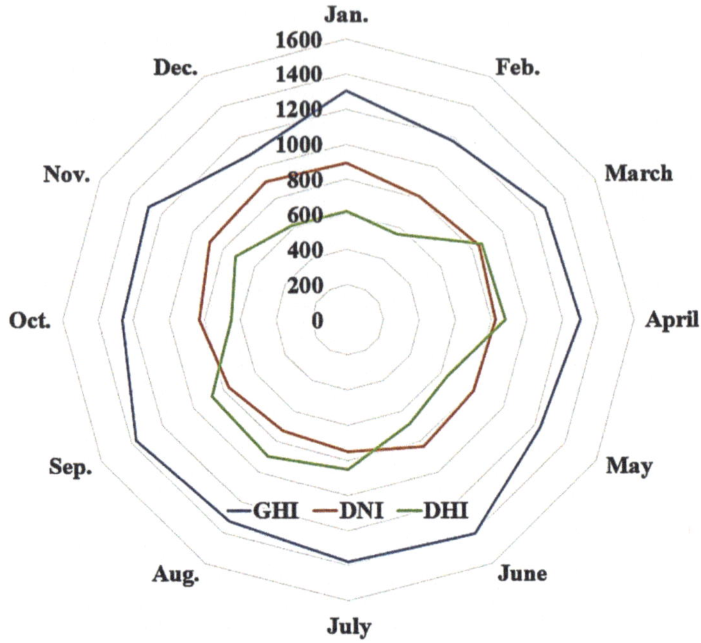

FIGURE 7.8 Historical maximum solar radiation of year 2015 for (A) GHI: Global horizontal irradiance [W/m^2], (B) DNI: Direct normal irradiance [W/m^2], and (C) DHI: Diffuse horizontal irradiance [W/m^2].

TABLE 7.1 Statics analysis of recorded dataset's variable of Temperature (°C).

Month	Minimum	Mean	Maximum	STD	Variance
January	19	25.6507	30.6	2.413765	5.826259
February	18.9	26.24307	31.3	2.485382	6.177122
March	20.7	27.35929	34.5	3.123979	9.759244
April	22.4	27.80584	34.3	2.596185	6.740177
May	22.3	27.67756	33.6	2.095094	4.38942
June	23	27.80154	32.4	1.819486	3.310529
July	23.2	28.06247	31.5	1.484181	2.202794
August	22.8	27.41071	31.7	1.662676	2.764491
September	23.4	26.88619	31.6	1.707266	2.914756
October	23.1	27.00814	31.5	1.924757	3.704691
November	22.7	27.51367	31.3	1.675828	2.808399
December	20.2	27.05633	31.3	2.044898	4.18161
Average	21.81	27.21	32.13	2.09	4.56

TABLE 7.2 Statics analysis of recorded dataset's variable of relative humidity (%).

Month	Minimum	Mean	Maximum	STD	Variance
January	44	77.08861	97	13.49376	182.0816
February	26	76.17773	96	11.71959	137.3487
March	20	75.95179	97	15.62088	244.0118
April	43	80.40263	97	13.05303	170.3816
May	54	84.60057	97	9.688328	93.8637
June	59	84.95416	97	7.81928	61.14114
July	64	83.08242	95	5.719084	32.70792
August	64	85.14238	95	6.760973	45.71076
September	63	86.42691	96	6.895918	47.55369
October	58	86.04175	97	8.644358	74.72493
November	64	83.97892	97	7.421851	55.08387
December	50	80.7023	97	9.299413	86.47907
Average	50.75	82.05	96.5	9.68	102.59

TABLE 7.3 Statics analysis of recorded dataset's variable of pressure (hPa).

Month	Minimum	Mean	Maximum	STD	Variance
January	1001	1006.091	1011	1.82202	3.319755
February	1002	1005.535	1011	1.701586	2.895395
March	1001	1005.173	1010	1.681943	2.828933
April	998	1003.54	1008	1.709452	2.922226
May	999	1001.962	1006	1.358595	1.845779
June	996	1001.075	1006	1.840213	3.386384
July	996	1000.606	1006	2.52531	6.377192
August	996	1001.936	1006	2.03283	4.132399
September	997	1002.631	1008	2.062705	4.254751
October	1000	1004.238	1007	1.409977	1.988035
November	1000	1003.762	1008	1.496247	2.238754
December	1002	1005.682	1011	1.802909	3.25048
Average	999	1003.52	1008.17	1.786	3.29

TABLE 7.4 Statics analysis of recorded dataset's variable of dew point temperature (°C).

Month	Minimum	Mean	Maximum	STD	Variance
January	15.6	21.144	25.6	1.787879	3.196511
February	9.7	21.58986	24.9	1.360483	1.850914
March	7.1	22.4152	25.6	1.851526	3.42815
April	18.8	23.94664	27.4	1.073994	1.153463
May	21	24.77644	27.7	0.902349	0.814234
June	21.6	25.0129	27.4	0.717343	0.514581
July	21.1	24.946	27.6	0.871081	0.758781
August	21.5	24.68613	27.3	0.698755	0.488259
September	22.2	24.41247	26.7	0.71967	0.517924
October	21.1	24.42462	27.3	0.654602	0.428504
November	20.2	24.55048	26.6	0.642756	0.413135
December	18.6	23.4047	26.4	1.198045	1.435311
Average	18.21	23.78	26.71	1.04	1.25

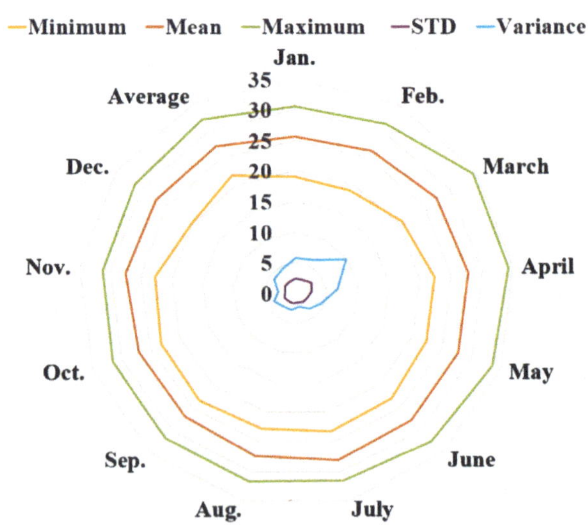

FIGURE 7.9 Air temperature (°C) analysis.

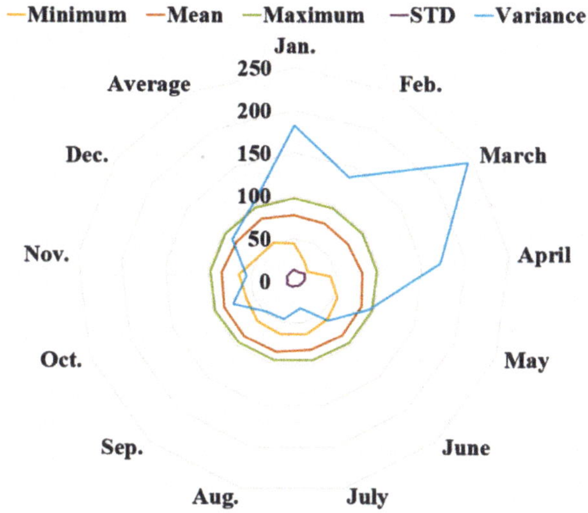

FIGURE 7.10 Relative humidity (%) analysis.

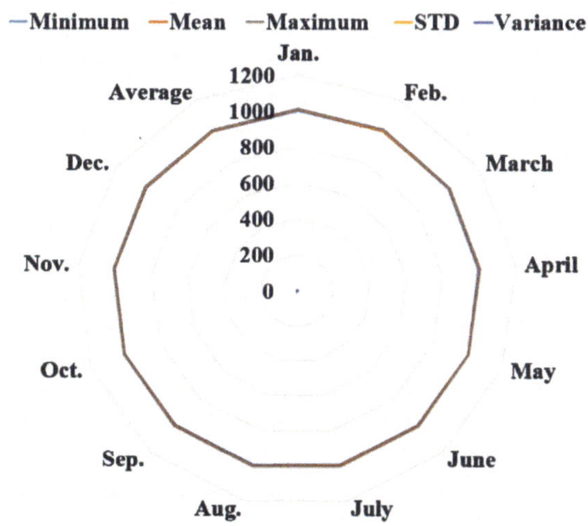

FIGURE 7.11 Pressure (hPa) analysis.

5 Structure of proposed model

In this section, the proposed DNN based solar radiation forecast (SRF) model is shown in Fig. 7.14, which shows the step-by-step procedure of the DNN model. Generally, the proposed method includes two sections: (1) data preprocessing and (2) SRF using DNN. The training and testing phases of DNN model are summarized into three basic steps: (1) initialization of parameters for DNN, (2)

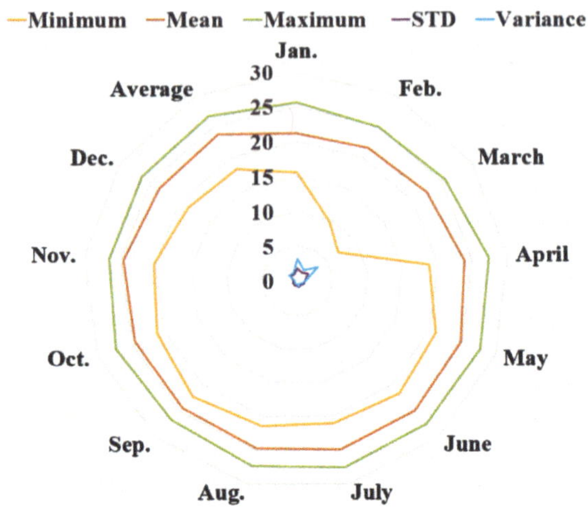

FIGURE 7.12 Dew point temperature (°C) analysis.

training the network with respect to iterations, and (3) testing and performance evaluation for benchmarking the performance.

Initially, DNN is developed using RNN (recurrent neural network) with some network configuration (L = depth, H = number of hidden unit, B = batch size, I = size of input, and O = size of output sequence). Then training iteration is performed until the forecasting loss in form of RMSE is in reduced form, which can be computed by Eq. (7.1):

$$Loss\Big|_{O_{forecasted},O_{measured}} = \sqrt{\frac{1}{B}\cdot\frac{1}{O}\cdot\sum_{i=1}^{B}\sum_{j=1}^{O}\left(O_{forecasted} - O_{measured}\right)^2} \tag{7.1}$$

5.1 Deep learning neural network

Generally, deep learning (DL) is the branch of machine learning methods. DL is the DNN, which is the upgraded/advanced version of MLP (multi-layer perceptron) neural network. The main difference in DNN is the use of activation function from the standard sigmoid function. The concept of DL has been proposed by McCulloch and Pitts in 1943 in form of the name "cybernetics." There are two main rationales for utilizing the DL, which are: (1) to learn highly nonlinear relationships and (2) to learn shared uncertainties.

Moreover, architecture development of DL includes CNN (convolutional deep neural networks), DRBM (deep restricted Boltzmann machines), DSA (deep sparse autoencoder), MLP, DRNN (deep recurrent neural networks), etc. The implementation of DRNN includes the stacking multiple RNN layers into a deep designed architecture. For the implementation point of view, reader may

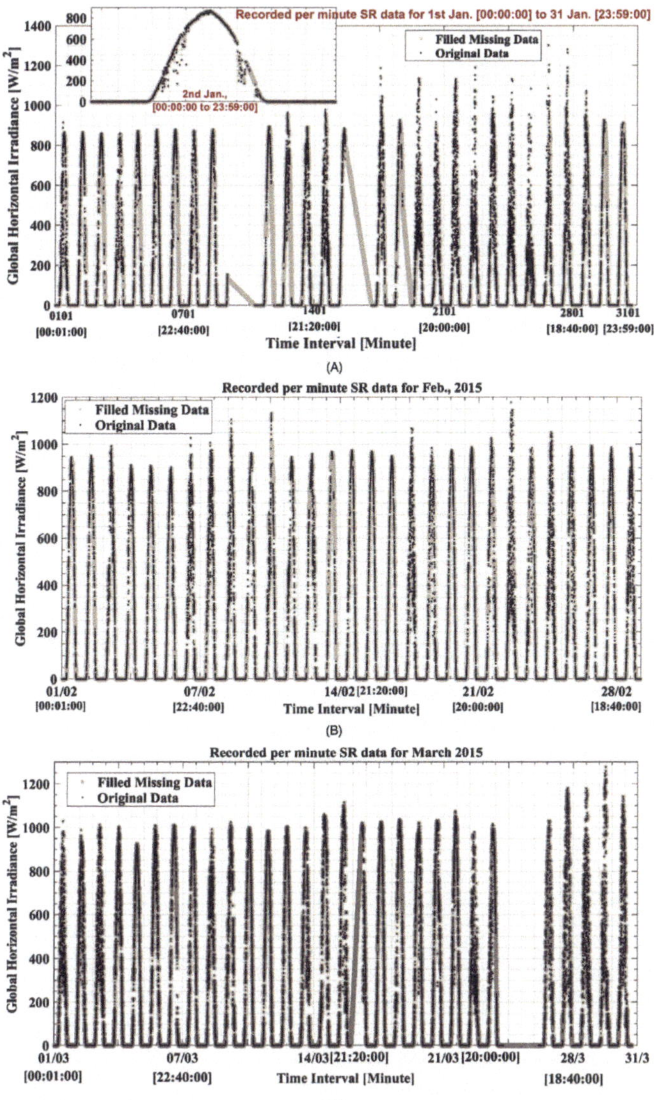

FIGURE 7.13 Monthly GHI representations for original historical data and filled missing value data.

refer Refs. [14,15] for evidence of benefit of building RNN with DL architecture. The detailed architecture with multiple layers (N-layers) for DRNN has been illustrated in Fig. 7.15.

In the Fig. 7.14, the RNN is used to map input data sample of I with respect to the output/target of O, and the learning is executed for each single time step

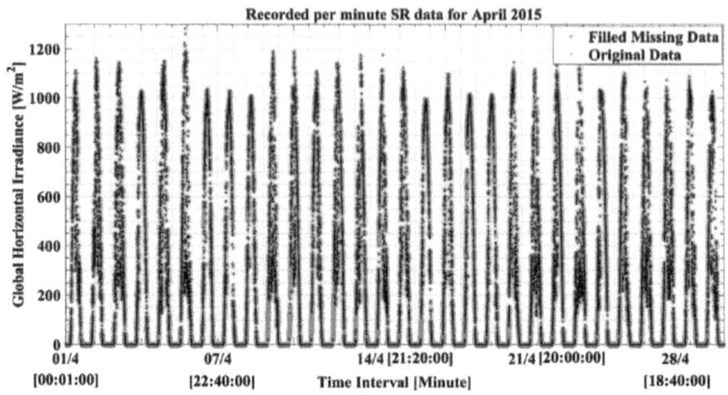

(D)

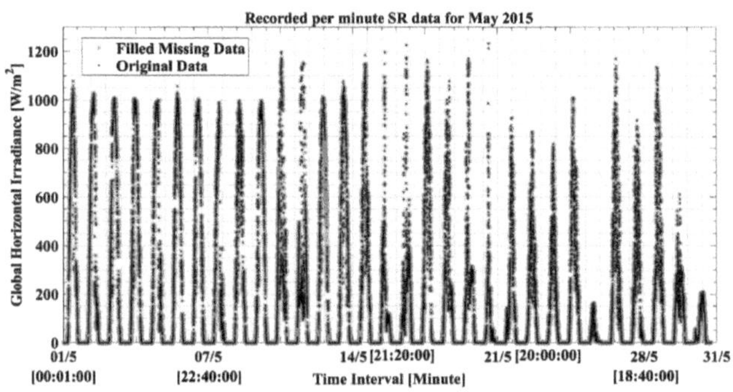

(E)

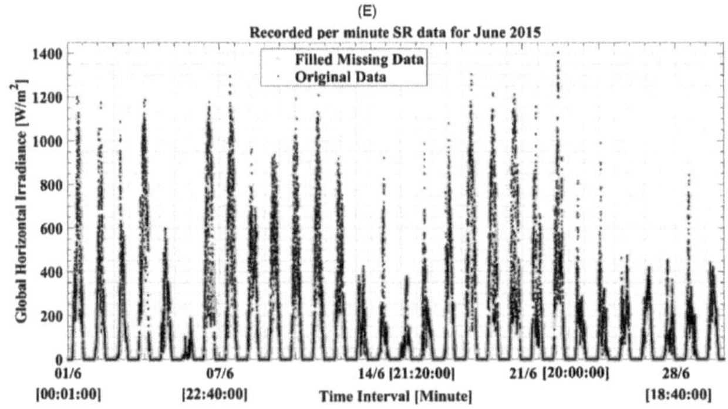

(F)

FIGURE 7.13 *(Cont.)*

varies from $t = 1$ to $t = \Gamma$. For t, the parameters of neuron at layer (lth) update its sharing states by using Eq. (7.2) as given [16]:

$$p_1^{(t)} = q_1 + \omega \cdot y_1^{(t-1)} + \eta_1 \cdot I^{(t)} \tag{7.2}$$

Where, $y_l^{(t)} = activation_{function\left(p_1^{(t)}\right)}$ for $l = 1, 2, ..., N$, q = bias

$$p_l^{(t)} = q_l + \omega_l . y_l^{(t-1)} + \eta_l . y_{l-1}^{(t)} \text{ for } l = 1, 2, ..., N \tag{7.3}$$

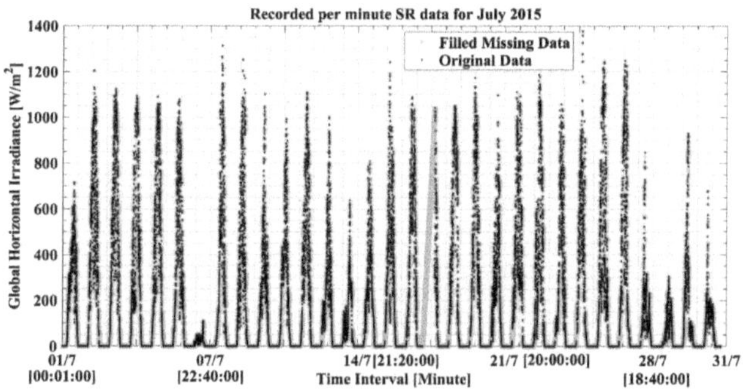

(G)

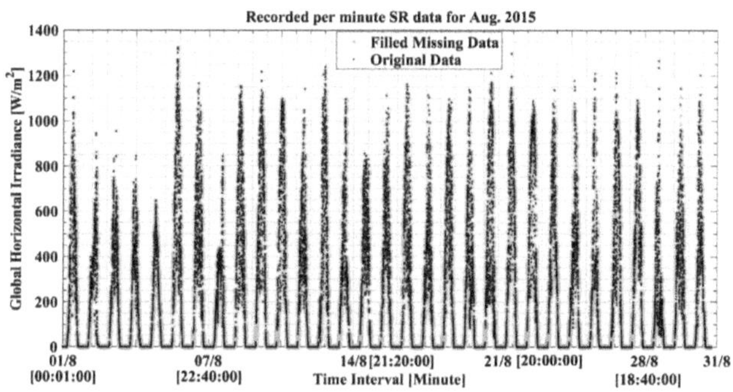

(H)

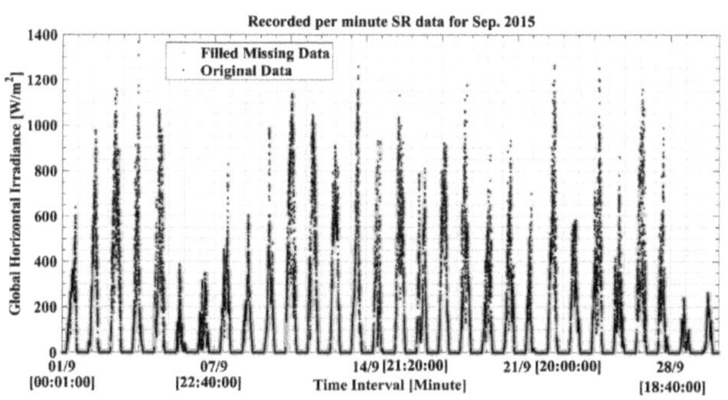

(I)

FIGURE 7.13 (*Cont.*)

$$O^{(t)} = q_N + \omega . y_N^{(t-1)} + \eta_N . y_N^{(t)} \tag{7.4}$$

$$\text{Loss, } f = loss\_function\left(O^{(t)}, O_{\text{target}}^{(t)}\right) \tag{7.5}$$

Where, $I^{(t)}$ = input data at t^{th} time-step, $O^{(t)}$ = predicted value, $O_{\text{target}}^{(t)}$ = measured value (output), $y_l^{(t)}$ = sharing states of lth network layer at t time, $p_l^{(t)}$ = input of lth layer at time step t, which comprises of three components: (1) bias (q), (2) sharing state $y_{l-1}^{(t)}$, and (3) t^{th} time step input $I^{(t)}$.

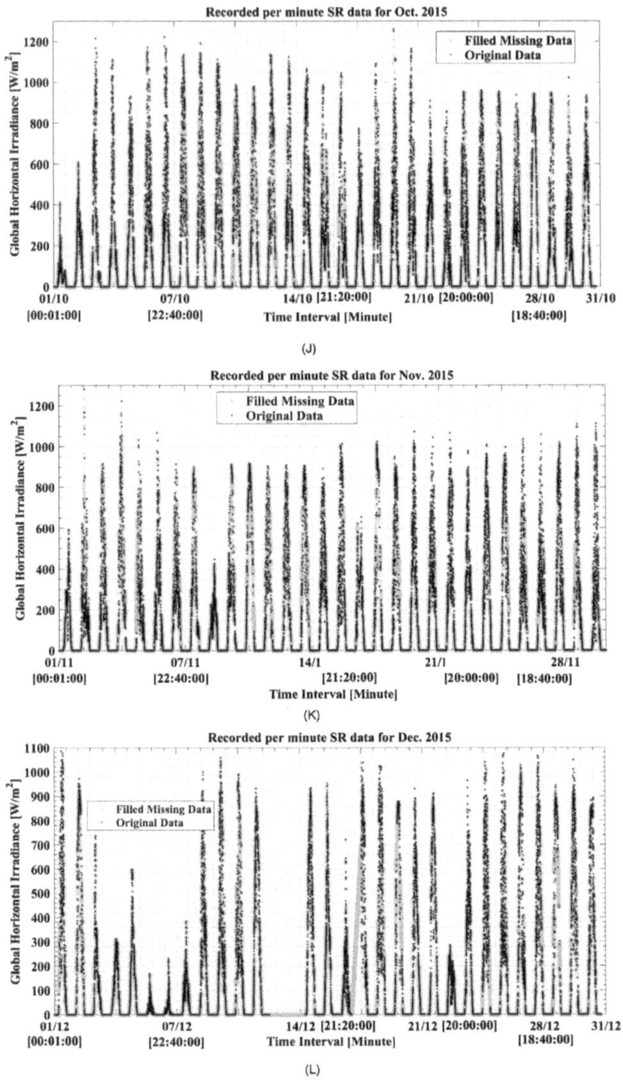

FIGURE 7.13 (*Cont.*)

5.2 Performance evaluation measures

The performance evaluation is the process to compute the model's characteristics in form of its performance, which shows the model's capability to STFP, which are MAE (mean absolute error), MAPE (mean absolute percentage error), and RMSE (root mean square error) used in this study. These performance indices are computed for the DNN model given as:

$$MAE = \frac{1}{N}\sum_{t=1}^{K}\left|\hat{f}(t) - f_h(t)\right| \tag{7.6}$$

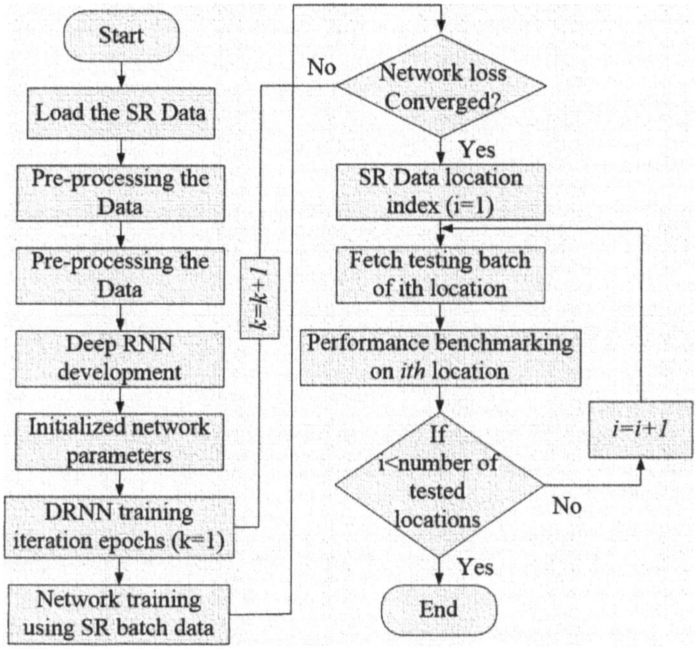

FIGURE 7.14 Proposed DNN based solar radiation forecast (SRF).

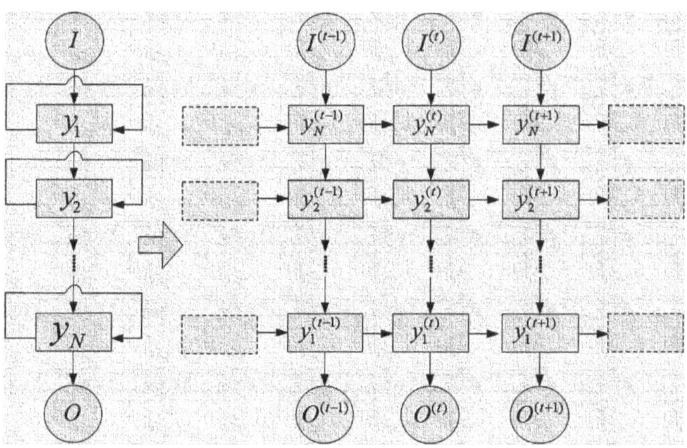

FIGURE 7.15 The detailed architecture with multiple layers (N-layers) of RNN for DRNN model.

$$MAPE = \left(\frac{1}{N} \sum_{t=1}^{K} \left| \frac{\hat{f}(t) - f_h(t)}{f_h(t)} \right| \right) \times 100 \tag{7.7}$$

$$RMSE = \sqrt{\frac{1}{N} \sum_{t=1}^{K} \left(\hat{f}(t) - f_h(t) \right)^2} \tag{7.8}$$

Where, $\hat{f}(t)$ and $f_h(t)$ are the forecasted and historical solar radiation at time interval k respectively, K is the total number of forecasted and historical data samples.

6 Results and discussion

This section represents the solar radiation (SR) forecasting obtained results with the help of proposed DNN for the different time interval of one-step ahead and which is forecasted upto step 4. For each step ahead forecasting of SR, the performance indices (MAE, MAPE and RMSE) are evaluated for each DNN model, which is shown in Table 7.5 for the one-step ahead forecasting. The Table 7.5 shows the outperformed results of proposed DNN model. Then based on these performance evaluation, DNN model was utilized to forecast the SR with two-step ahead, three-step ahead, and four-step ahead with the same dataset with different time interval. In the multi-step forecasting, following assumptions were considered: one-step (1-min), two-step (2-min), three-step (3-min), and four-step (4-min) ahead forecasting. The learning curve for training phase of DNN model is represented in Fig. 7.16, in which dark blue and dark orange color lines have represented the training (smoothed) RMS and loss function of the model, respectively. Similarly, light blue and light orange color lines are representing the training (output) RMS and loss function of the model, respectively, which are very negligible in this case.

Fig. 7.17 represents the forecasting behavior and performance indices obtained with the proposed DNN for overall time period of records. Presented results shows that proposed DNN method outperforms for all cases of multi-step, which presents the acceptable value of MAE, RMSE, and MAPE error. The capability of the proposed DNN method is demonstrated for its effectiveness by

TABLE 7.5 One-step ahead SR forecasting performance indices representation.

		MAE		RMSE		MAPE (%)	
		DNN	ANN	DNN	ANN	DNN	ANN
2015 Dataset	Training phase	0.302	3.82	1.83	4.50	2.95	13.80
	Testing phase	0.587	4.50	2.901	6.25	7.821	17.50
2016 Dataset	Testing phase	1.140	5.24	5.515	8.32	13.52	16.10
2017 Dataset	Testing phase	1.312	5.00	5.942	8.40	14.53	18.25

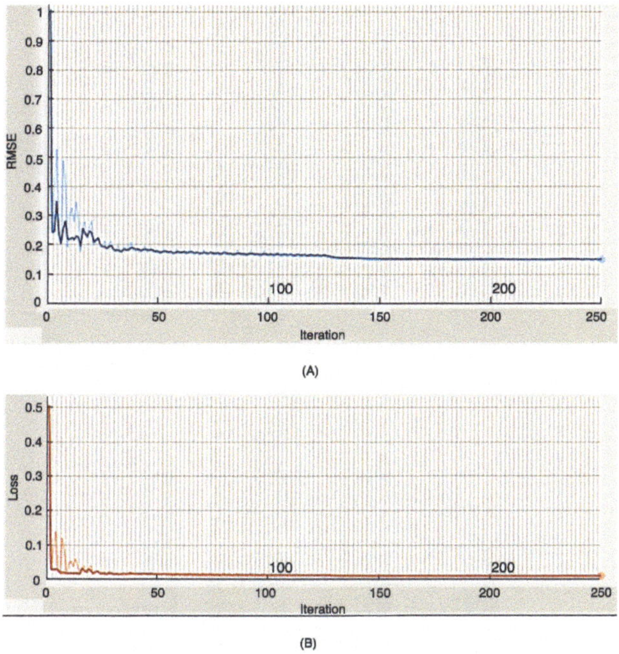

FIGURE 7.16 **DNN Learning curve for training phase of SRF.** (A) RMSE and (B) Loss function.

using Port Blair historical dataset from the year 2015 to 17 and the testing phase results for each month, which shows the high level of acceptability of the proposed DNN model.

7 Conclusion

A new advanced DNN approach is proposed for multi-step ahead SRF and its performance is validated and demonstrated by using Indian city of Port Blair for short-term forecasting of SR. This NN model is capable of solving the linear as well as non-linear problems. So, SRF is a highly non-linear problem which is short-term forecast (per minute), which is handled by proposed DNN method in an efficient way with less computational burden. The multi-step ahead forecasting is performed recursively, using historical datasets one-step (1-min), two-step (2-min), three-step (3-min), and four-step (4-min) ahead forecasting, and this procedure is repeated until a time step of 4 min ahead. The approach is demonstrated using real meteorological data from Port Blair, a city of India. The demonstration of the proposed DNN model is accomplished comparing the obtained results with ANN model. The demonstrated results showed that the proposed DNN model for SRF outperformed with respect to the all criterion (MAPE, RMSE, and MAE) of ANN model for one-step ahead forecasting results.

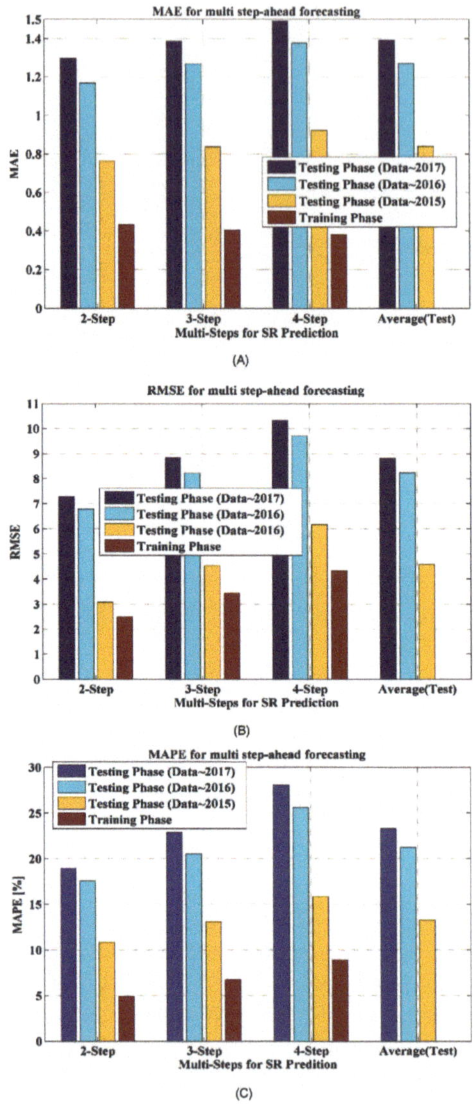

FIGURE 7.17 The representation of SRF model performance indices for multi-step forecasting. (A) MAE, (B) RMSE, and (C) MAPE.

References

[1] H. Ritchie, M. Roser, Renewable Energy. Available from: https://ourworldindata.org/renewable-energy.

[2] T.V. Ramachandra, R. Jain, G Krishnadas, Hotspots of solar potential in India, Renew. Sustain. Energy Rev. 15 (2011) 3178–3186.

[3] M.S. Soni, Gakkhar N, Techno-economic parametric assessment of solar power in India: A survey, Renew. Sustain. Energy Rev. 40 (2014) 326–334.

[4] V. Khare, S. Nema, Baredar P, Status of solar wind renewable energy in India, Renew. Sustain. Energy Rev. 27 (2013) 1–10.
[5] http://pib.nic.in/newsite/erelease.aspx?relid=83632. Accessed 17.01.2014.
[6] T. Khatib, A. Mohamed, Sopian K, A review of photovoltaic systems size optimization techniques, Renew. Sustain. Energy Rev. 22 (2013) 454–465.
[7] H. Lu, W. Zhaob, Effects of particle sizes and tilt angles on dust deposition characteristics of a ground-mounted solar photovoltaic system, Appl. Energy 220 (2018) 514–526.
[8] P. Sawicka-Chudy, M. Sibiński, M. Cholewa, Pawełek R, Comparison of solar tracking and fixed-tilt photovoltaic modules in Lodz, J. Sol. Energy Eng. 140 (2) (2018) 1–6.
[9] A.K. Yadav, H. Malik, S.S. Chandel, Selection of most relevant input parameters using WEKA for artificial neural network based solar radiation prediction models, Renew. Sustain. Energy Rev. 31 (2014) 509–519.
[10] A.K. Yadav, et al., Application of Rapid Miner In ANN Based Prediction Of Solar Radiation for Assessment of Solar Energy Resource Potential of 76 Sites in Northwestern India, Renew. Sustain. Energy Rev. 52 (2015) 1093–1106, doi: 10.1016/j.rser.2015.07.156.
[11] A.K. Yadav, et al., Comparison of Different Artificial Neural Network Techniques in Prediction of Solar Radiation for Power Generation Using Different Combinations of Meterological Variables, In Proc. IEEE Int. Conf. on Power Electronics, Drives and Energy Systems (PEDES-2014) (2014) 1–5, doi: 10.1109/PEDES.2014.7042063.
[12] S.S. Chandel, et al., ANN Based Prediction of Daily Global Solar Radiation for Photovoltaics Applications, In Proc. IEEE India Annual Conference (INDICON) (2015) 1–5, doi: 10.1109/INDICON.2015.7443186.
[13] S. Garg, Long-Term Solar Irradiance Forecast Using Artificial Neural Network: Application for Performance Prediction of Indian Cities, Applications of Artificial Intelligence Techniques in Engineering, Advances in Intelligent Systems and Computing 697 (2018) 285–293, https://doi.org/10.1007/978-981-13-1822-1_26.
[14] A. Graves, A.-R. Mohamed, G. Hinton, Speech recognition with deep recurrent neural networks, in: Proceedings of IEEE International Conference on Acoustic, Speech and Signal Processing, Vancouver, BC, Canada, 2013, pp. 6645–6649.
[15] R. Pascanu, C. Gulcehre, K. Cho, Y. Bengio, How to construct deep recurrent neural networks, arXiv preprint arXiv:1312.6026, 2013.
[16] I. Goodfellow, Y. Bengio, A. Courville, Deep Learning, MIT Press, Cambridge, MA, USA, (2016) http://www.iro.umontreal.ca/~bengioy/dlbook.

Chapter 8

Intelligent Data Analytics for Wind Speed Forecasting for Wind Power Production Using Long Short-Term Memory (LSTM) Network

1 Introduction

Nowadays, modernization of the industry and smart city, the requirement of energy has been increased drastically. In order to meet the power consumption limit, utilities are trying to enhance utilization of the renewable resources. According to cited work in Ref. [1], around 65% contribution of the total global power addition in 2018 was only form renewable energy source (RES). The generation of wind energy is increasing day-by-day as in 1965, generated wind power was zero but now 3.63, 31.42, 341.61, 1127.99, and 1269.95 TWh in the year of 1990, 2000, 2010, 2017, and 2018, respectively [2], which shows the great sign of utilization reduction in traditional biofuel-based energy consumption. Developing nations are rapidly shifting toward the renewable energy, and India has become one of the leaders in this sector.

According to the MNRE (Ministry of New and Renewable Energy Resources), India creates an ambitious objective of 175 GW of renewable energy generation by 2022 [3]. According to the available wind power potential and resources, India is the fourth largest economy in the world with 32.7 GW of installed wind power capacity. As per the survey and analysis conducted in the year 2018 by NIWE (National Institute of Wind Energy), it is estimated that the total wind power potential of Indian sub-continent is about 302 GW [4].

Based on time-scale, WSFP (wind speed forecasting and prediction) is categorized into four categories: (1) very-short term (VST), (2) short-term (ST), (3) medium-term (MT), and (4) long-term (LT) [5,6]. The VST, ST, MT, and LT WSFP are based on the time interval in order of "seconds to few minutes," "(30 min–24 h)," "(1 day–1 week)," and "(1 week–years)," respectively. Moreover, the different methods of WSFP are represented in the available literature and are

Intelligent Data-Analytics for Condition Monitoring. http://dx.doi.org/10.1016/B978-0-323-85510-5.00008-9

mainly categorized into four types: (1) physical methods (PM), (2) statistical methods (SM), (3) artificial intelligence based methods (AIM), and (4) hybrid methods (HM).

In the available literature, there are numerous research papers, inventions and copyrights related to the hybrid method of WSFP. For instance, Lili Wang and coworkers [7] developed a HM comprising ELM-ARIMA methods for one step ahead WSF. Zhongye and coworkers [8] proposed a HM based on ARIMA method based on Kalman filter (KF) by using three different location's historical data. Alencar and coworkers [5] proposed a HM based on SARIMA with ANN for multi-step ahead WSF for two stations of Brazil. Jing Shi And coworkers [9] developed two HM using ARIMA-ANN and ARIMA-SVM for multi-step ahead WSF using historical data from Colarado, USA. Erasmo Cadenas and coworkers [10] developed a HM using ARIMA and ANN for ST WSF of three sites of California. M. Yesilbudak and colleagues [11] presented a K-Nearest Neighbor (kNN) algorithm for WSFP four input variables (i.e., wind direction, air temperature, atmospheric pressure, and relative humidity). Jianzhou Wang and coworkers [12] presented a detailed review on WSF, including various application of wind energy, a detailed explanation of various time horizons for wind speed forecasting and methods of WSF.

The WS range between 4 and 35 m/s is the productive WS, which can be used for electricity generation by a particular wind turbine. Currently, the amount of power generated by the WS has been raised drastically. The introduced WE has expanded by almost 200% in the vicinity of 2005–09.

Nowadays, WE is the most growing resource in the world. WE is needed for assessment of WS at different times and is of stochastic nature. WS is converted into power using wind turbines. The wind power is a function of WS. The accurate power prediction of wind farm output requires prediction of WS. Prediction of WS is important for wind farms units' maintenance, optimal power flow between conventional units and wind farms, electricity marketing bidding, power system generators scheduling, energy reserve, storages planning, and scheduling. Therefore, predictions of WS performed by various researchers are discussed in subsequent sections [13].

The main contribution of this paper is to present a novel LSTM approach based on a real-time measured historical dataset of 2015–17 for WSFP. Furthermore the WS is forecasted using a developed model for Indian sites where there is no installed wind speed measuring instrument. The predicted WS can be utilized for wind resource assessment for installation of wind turbines to generate the power.

The formulation of this chapter is presented into five sections, which includes the introduction in Section 1 and intelligent data analysis for WSFP in Section 2. Section 3 represents the proposed framework, which includes the following sub-sections: (1) proposed approach formation in Section 3.1, (2) dataset collection for the study in Section 3.2, (3) features extraction in Section 3.3, (4) features selection in Section 3.4, (5) design of LSTM network and its modified version in Section 3.5, and (6) performance measure indices in Section 3.6. Results demonstration and its discussion are represented in Section 4 and finally conclusion is shown in Section 5.

2 Intelligent data analysis for WSFP

The WSFP is a part of RES (renewable energy sources) forecasting and prediction. Although, RES forecasting and prediction are the mature technologies with operational tools and services used by different users. However, there are several research gaps and bottlenecks in the models and applications encouraging significant research worldwide. The main objective of this study is to present the challenges and innovative research toward the intelligent data analytics for RES forecasting and prediction using machine learning.

There are different RES analytics tools and techniques. Renewable energy decision can be categorized by seven-key analytical approaches using the data. These analytical approaches are outlined as: (1) Early-stage mapping and visualization (ESMV), (2) Generator performance modeling (GPM), (3) Technical potential analysis (TPA), (4) Supply curve modeling (SCM), (5) Economic potential analysis (EPA), (6) Capacity expansion modeling (CEM), and (7) Production cost modeling (PCM). The feasibility and utilization of these analytical techniques depend on the required dataset as well as the level of analytical complexity, which is shown in Fig. 8.1. In Fig. 8.1, the early-stage RE development meaning is related to decision making supported by high level visualization of RES. Similarly, advanced-stage RE development is related to decision making

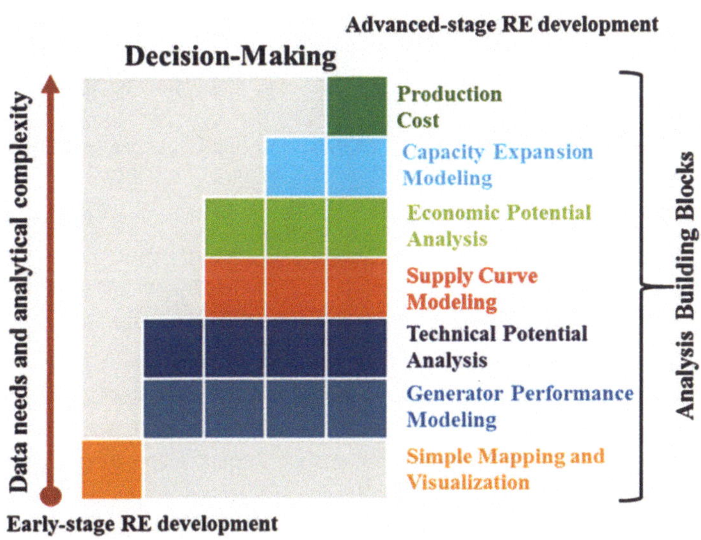

FIGURE 8.1 **Analysis building blocks to support RE decisions.**

supported by robust analysis with several analytical building blocks such as capacity expansion modeling, etc. The blocks shown in the Fig. 8.1 represent the level of data volume required for the analysis by individual approach. So, *x*-axis of the Fig. 8.1 represents the level of volume of the required dataset whereas, *y*-axis represents the level of the complexity of the approach, which varies from low to high (bottom to top side). For example, first block (in orange color) at the bottom side of Fig. 8.1 shows that analytical approach #1 (i.e., ESMV) requires less amount of data for the study whereas, its level of complexity is also very low. There is also one block (in dark green color) at the top side of the Fig. 8.1, which shows that analytical approach (PCM) has very high complexity in analysis but it requires less amount of dataset similar to ESMV.

There are different types of tools for each type of analytical approach, which are listed in the Table 8.1. These tools can be requested from respective organization for the academic purpose trough a proper channel. Moreover, the reader needs a database for the further validation of the developed model using earlier-mentioned tools and software in Table 8.1. To overcome such problem, few online available databases have been summarized in the Table 8.2. There

TABLE 8.1 Available tools for analytical approaches of RES.

S. no.	Analytical approach	Available tools for study	URL
1	ESMV	IRENA Global Energy Atlas	http://irena.masdar.ac.ae/
		Lawrence Berkeley National Laboratory MapRE	http://mapre.lbl.gov/
		Renewable Energy Data Explorer	https://www.re-explorer.org/
		World Bank Group Global Solar Atlas	http://globalsolaratlas.info/
		World Bank Group Global Wind Atlas	https://globalwindatlas.info/
2	GPM	Systems Advisor Model (SAM)	https://sam.nrel.gov/
		PVSyst	http://www.pvsyst.com/
		HOMER	http://www.homerenergy.com/
		Windographer	https://www.windographer.com/
3	TPA	Renewable Energy Data Explorer	https://www.re-explorer.org/
		Renewable Energy Potential (reV) model	https://www.nrel.gov/gis/modeling.html
		Lawrence Berkley National Laboratory MapRE (ArcGIS Tools)	http://mapre.lbl.gov/gis-tools/

TABLE 8.1 Available tools for analytical approaches of RES. (*Cont.*)

S. no.	Analytical approach	Available tools for study	URL
4	SCM	Renewable Energy Potential Model (reV)	https://www.nrel.gov/gis/modeling.html
		Lawrence Berkeley National Lab MapRE	http://mapre.lbl.gov/
5	EPA	RED-E platform - for ASEAN countries	
		Others tool	http://www.nrel.gov/gis/re_econ_potential.html
6	CEM	Resource Planning Model (RPM)	http://www.nrel.gov/analysis/models_rpm.html
		Regional Energy Deployment System (ReEDS)	http://www.nrel.gov/analysis/reeds/
		PLEXOS-LT	https://energyexemplar.com/solutions/plexos/
7	PCM	PLEXOS	https://hpc.nrel.gov/projects/latest/fy160060
		GridView	

TABLE 8.2 Summary of online available datasets.

S. no.	The online available datasets	Websites
1	The National Oceanic and Atmospheric Administration (NOAA)	http://www.noaa.gov
2	US Department of Energy, NREL	https://rredc.nrel.gov/solar/new_data/confrrm/bs/
3	US Department of Commerce, Earth System Research Laboratory	https://www.esrl.noaa.gov/gmd/grad/surfrad/
4	National Solar Resource Database	https://maps.nrel.gov/nsrdb-viewer
5	The Renewable Energy (RE) Explorer and the RE Data Explorer	http://www.re-explorer.org/
6	Lawrence Berkeley National Laboratory Map Re	http://mapre.lbl.gov/

(*Continued*)

TABLE 8.2 Summary of online available datasets. (*Cont.*)

S. no.	The online available datasets	Websites
7	Danish Technical University Global Wind Atlas	https://globalwindatlas.info/
8	The World Bank Group Global Solar Atlas	http://globalsolaratlas.info/
9	Cloud Images And Inverter Data	http://www.findlatitudeandlongitude.com
10	Sky Images Data	http://solar.research.nicta.com.au/
11	PV Measurements	http://www.PVOutput.org
12	Australian Bureau of Meteorology's (BoM)	http://reg.bom.gov.au/climate/reg/oneminsolar/index.shtml
13	IEEE DataPort	https://ieee-dataport.org/
14	NASA SRB	http://gewex-srb.larc.nasa.gov/
15	DLR-ISIS	http://www.pa.op.dlr.de/ISIS/
16	HelioClim	http://www.soda-is.com/eng/helioclim/
17	SOLEMI	http://wdc.dlr.de/dataproducts/SERVICES/SOLARENERGY/
18	SolarGIS	http://solargis.info/
19	EnMetSol	https://www.unioldenburg.de/en/physics/research/ehf/energiemeteorology/enmetsol/
20	IrSOLaV	http://irsolav.com/
21	CM SAF (SARAH)	http://www.cmsaf.eu/
22	SolarAnywhere	http://www.solaranywhere.com/
23	CAMS	http://atmosphere.copernicus.eu/catalogue/
24	PVGIS	http://re.jrc.ec.europa.eu/pvgis/
25	Vaisala	http://www.vaisala.com
26	Australian Bureau of Meteorology	http://www.bom.gov.au/climate/dataservices/solarinformation.shtml

are some paid and some are open access database. Regarding the utilization of the database, reader needs to be concerned with appropriate authority of the respective database for the purpose of intelligent data analytics of RES applications.

3 Proposed framework formation

The proposed framework formation for WSFP includes the following operations: (1) proposed approach formation, (2) historical dataset collection, (3) data preprocessing, (4) feature extraction, (5) most relevant feature selection, (6) design of LSTM network, and (7) performance measure indices. All these operations have been presented further.

3.1 Proposed approach formation

For the WSFP, the proposed approach formation has been shown in Fig. 8.2, which includes three main parts such as: (1) Data processing part, (2) LSTM model design and training part, and (3) Multi-step testing and adaptability

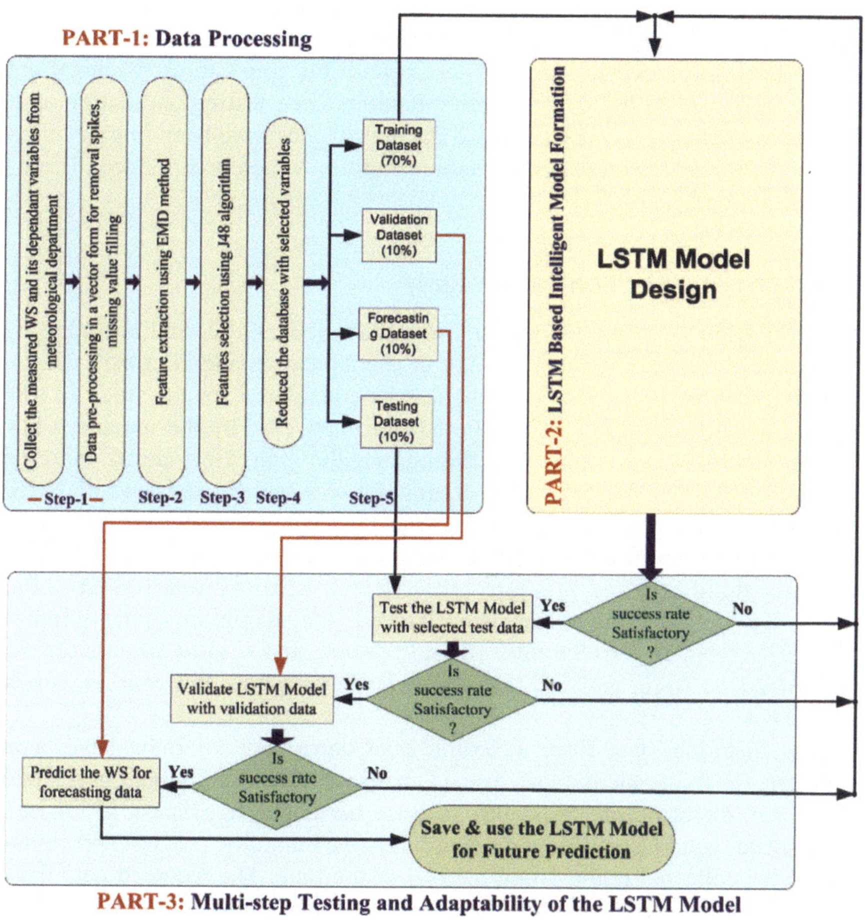

FIGURE 8.2 Proposed approach for WSFP.

verification part. First two parts (PART-1 and PART-2) include the six basic steps, which are represented as follows: Part-1 includes five steps and Part-2 includes one step. The dataset collection, data preprocessing (spikes removal and missing value filling), feature extraction, most relevant feature/variable selection, data reduction, and four different data files (i.e., training-70%, testing-10%, validation-10%, and forecasting files-10%) formation are included in the Part-1. In the first step, a recorded meteorological historical dataset from the sites has been collected with 13 variables. Now this dataset is arranged in a vector form which is used in the attribute selection technique as an input variable. After selection of most appropriate input variables, for different data files have been prepared. The developed all four data files are not similar to each other. The training data file is used for LSTM models design and developed LSTM models learns the WS prediction property based on per minute recorded dataset of Indian city. After training phase, testing phase is performed to test the trained model performance. If testing phase performance is acceptable then validation data file is used to test the model again. Therefore, trained model is tested multiple times to cross check the model performance by using different datasets (i.e., testing, validation, and forecasting dataset). LSTM and mLSTM approaches have been implemented with and without selected input variable condition which is explained in coming subsections.

3.2 Dataset collection for the study

The Port Blair (latitude [°N] 11.61, longitude [°E] 92.72, and altitude [m] 65) site in the south Andaman Island of the Andaman and Nicobar Islands, which is located in the Bay of Bengal, was considered for the study in this chapter as shown in the Fig. 8.3. Port Blair is covered by the sea and it has huge potential of solar potential throughout the year. Here mean value of WS varies from 3.75–1.61, 3.7–1.8, and 4.1–1.9 m/s for the year of 2015, 2016, and 2017, respectively, throughout the year. Moreover, it has tropical climate with significant rainfall throughout the year. The average temperature, humidity, and pressure are 27.21°C, 82.05%, and 1003.52 hPa, respectively. According to historical record, the maximum and minimum WS varies from 10.5 to 5.9 m/s, 18.8 to 5.7 m/s, and 11.3 to 5.8 m/s for the year of 2015, 2016, and 2017, respectively, throughout the year as shown in Fig. 8.4.

The historical data from meteorological department of India have been collected for the academic use. The 11 main variables (i.e., global horizontal irradiance, direct normal irradiance, diffuse horizontal irradiance, irradiance, wind speed, wind direction, temperature, relative humidity, pressure, precipitation) were collected at the 1-min interval of the time. The recorded data from January 1st, 2015 to December 31st, 2017 were used for the study. The further

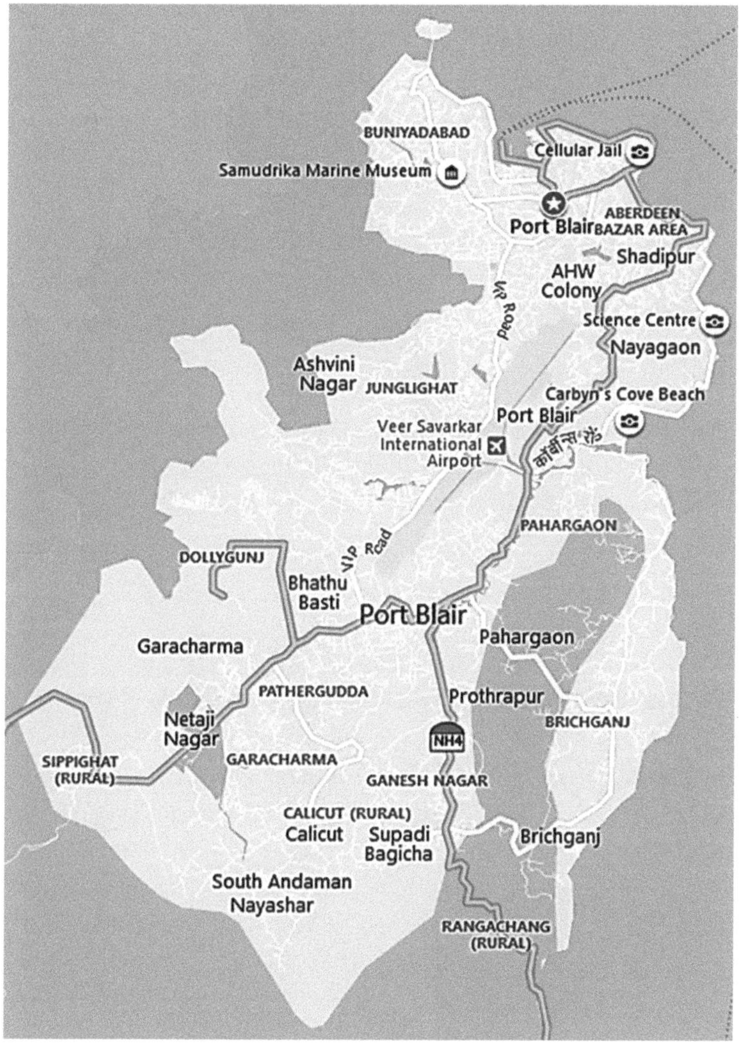

FIGURE 8.3 Location of Port Blair (shown in *dark white color*).

statistical analysis of the recorded data sets are shown in Tables 8.3–8.8 and Figs. 8.5–8.10, which are minimum value, maximum value, mean, standard deviation (STD), and variance (VAR) for the year of 2015.

The historical data of WSFP are recorded per minute basis and assumed time duration for the recording of a day is [00:00:00] to [23:59:00] for each day. Generally, there are some missing values and spikes in the historical time series data due to unwanted weather condition and/or instrumental/operational/technical errors. Therefore, data preprocessing technique is implemented and purifies

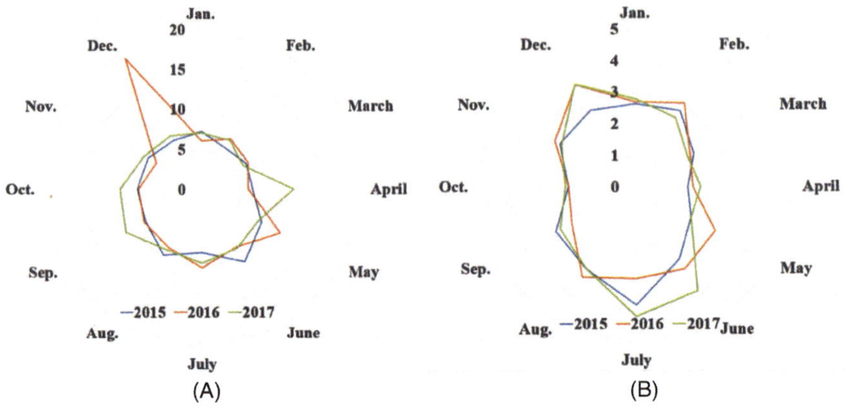

FIGURE 8.4 Historical wind speed of year 2015–17 for (A) maximum WS [m/s], (B) mean WS [m/s].

TABLE 8.3 Statistical analysis of recorded dataset's variable of wind speed (WS) (in m/s).

Month	Minimum	Mean	Maximum	STD	Variance
January	0	2.609154	7.2	1.205736	1.453799
February	0	2.774161	5.9	1.199777	1.439466
March	0	2.072197	6.3	1.131988	1.281398
April	0	1.606616	6.2	0.985084	0.97039
May	0	1.971066	8.4	1.079966	1.166327
June	0	2.67358	10.5	1.153815	1.331289
July	0	3.746041	8	1.095808	1.200794
August	0	3.042672	9.6	1.172948	1.375807
September	0	2.860911	8.1	1.092588	1.19375
October	0	2.074135	8	0.997377	0.994762
November	0	2.700758	7.7	1.114924	1.243056
December	0	2.782262	7	1.104303	1.219485
Average	0	2.576129	7.741667	1.111193	1.239194

TABLE 8.4 Statistical analysis of recorded dataset's variable of maximum horizontal wind speed (maxws) (in m/s).

Month	Minimum	Mean	Maximum	STD	Variance
January	0	2.609094	12.5	1.849951	3.42232
February	0	2.774161	9.2	1.844664	3.402786
March	0	2.072173	10.2	1.724281	2.973144
April	0	1.606616	9.1	1.466405	2.150343
May	0	1.971053	14.9	1.700566	2.891924
June	0	2.67358	14.8	1.857947	3.451966
July	0	3.746041	13.4	1.750307	3.063576
August	0	3.042685	15.4	1.866261	3.48293
September	0	2.860911	12.6	1.696241	2.877235
October	0	2.074162	12.3	1.518341	2.305359
November	0	2.700758	13.6	1.715263	2.942126
December	0	2.782256	11	1.696406	2.877794
Average		2.576124	12.41667	1.723886	2.986792

TABLE 8.5 Statistical analysis of wind direction (WD) (in degrees).

Month	Minimum	Mean	Maximum	STD	Variance
January	11	113.3685	346	71.48592	3.422302
February	13	93.74602	350	70.6211	3.402786
March	8	135.6592	352	84.36559	2.973185
April	6	151.376	351	80.21527	2.150343
May	11	206.8854	352	55.52652	2.891981
June	10	212.4206	349	39.06411	3.451966
July	17	225.2175	337	28.81539	3.063576
August	9	216.1727	343	38.1182	3.483016
September	12	228.8584	345	34.6028	2.877235
October	9	160.5083	348	77.32438	2.305377
November	9	97.44697	348	57.85684	2.942126
December	13	95.594	351	69.04584	2.877857
Average	10.66667	161.4378	347.6667	58.92016	2.986813

TABLE 8.6 Statistical analysis of recorded dataset's variable of relative humidity (RH) (in %).

Month	Minimum	Mean	Maximum	STD	Variance
January	44	77.08861	97	13.49376	182.0816
February	26	76.17773	96	11.71959	137.3487
March	20	75.95179	97	15.62088	244.0118
April	43	80.40263	97	13.05303	170.3816
May	54	84.60057	97	9.688328	93.8637
June	59	84.95416	97	7.81928	61.14114
July	64	83.08242	95	5.719084	32.70792
August	64	85.14238	95	6.760973	45.71076
September	63	86.42691	96	6.895918	47.55369
October	58	86.04175	97	8.644358	74.72493
November	64	83.97892	97	7.421851	55.08387
December	50	80.7023	97	9.299413	86.47907
Average	50.75	82.04585	96.5	9.678039	102.5907

TABLE 8.7 Statistical analysis of recorded dataset's variable of precipitation (precip) (in mm).

Month	Minimum	Mean	Maximum	STD	Variance
January	0	0.003542	1.4	0.04589	0.002106
February	0	0.000515	2	0.018352	0.000337
March	0	6.78E-06	0.1	0.000824	6.78E-07
April	0	0.001066	1.8	0.025556	0.000653
May	0	0.010166	2.9	0.092009	0.008466
June	0	0.009944	2.2	0.065339	0.004269
July	0	0.004096	1.8	0.045026	0.002027
August	0	0.010968	2.9	0.091203	0.008318
September	0	0.011208	2	0.077826	0.006057
October	0	0.006919	2.1	0.06358	0.004042
November	0	0.004293	2.1	0.052321	0.002737
December	0	0.00172	1.7	0.03598	0.001295
Average	0	0.00537	1.916667	0.051159	0.003359

TABLE 8.8 Statistical analysis of recorded dataset's variable of wet bulb temperature (wb) (in °C).

Month	Minimum	Mean	Maximum	STD	Variance
January	18.6	22.53484	26.1	1.362475	1.856339
February	18.4	22.97415	26	1.261432	1.591212
March	19.8	23.89298	27.3	1.349863	1.82213
April	18.8	23.94664	27.4	1.073994	1.153463
May	21.6	25.52807	28.5	0.961028	0.923575
June	22	25.72236	28.1	0.851575	0.725181
July	22.2	25.73627	28.1	0.911165	0.830221
August	22.1	25.38642	27.9	0.835457	0.697988
September	22.6	25.05392	27.4	0.868826	0.754858
October	22.4	25.09879	27.6	0.825666	0.681725
November	22.1	25.3196	27.6	0.76983	0.592639
December	19.8	24.40821	27.5	1.164841	1.356855
Average	20.86667	24.63352	27.45833	1.019679	1.082182

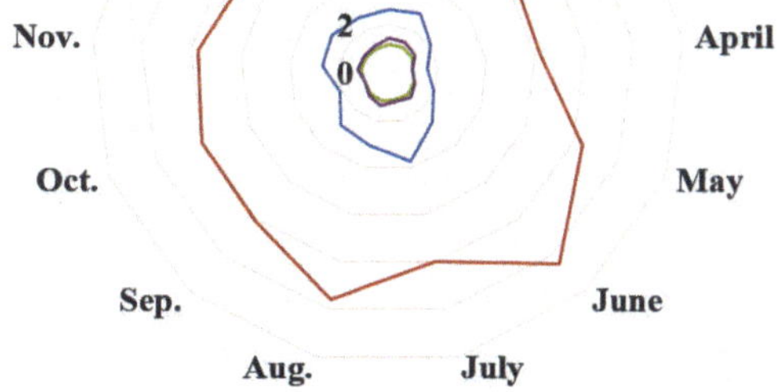

FIGURE 8.5 WS (m/s) analysis.

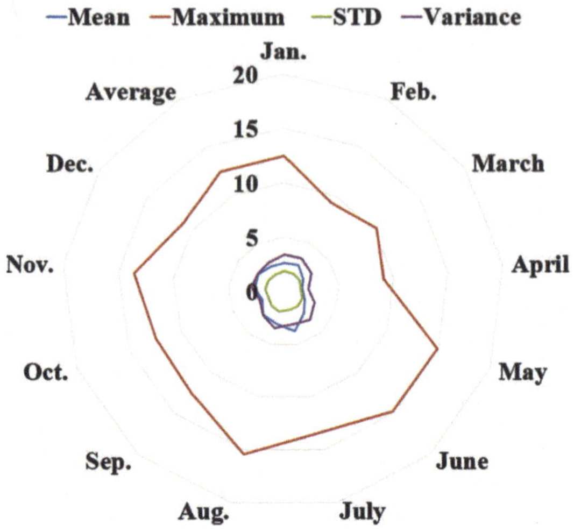

FIGURE 8.6 Maximum horizontal wind speed (maxws) (in m/s).

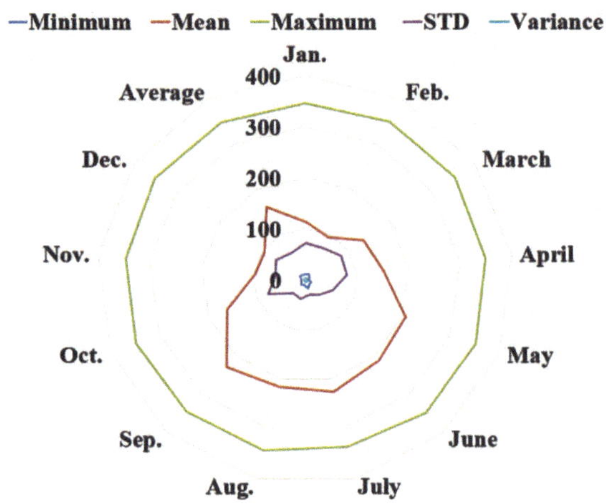

FIGURE 8.7 Wind direction (WD) (in degree) analysis.

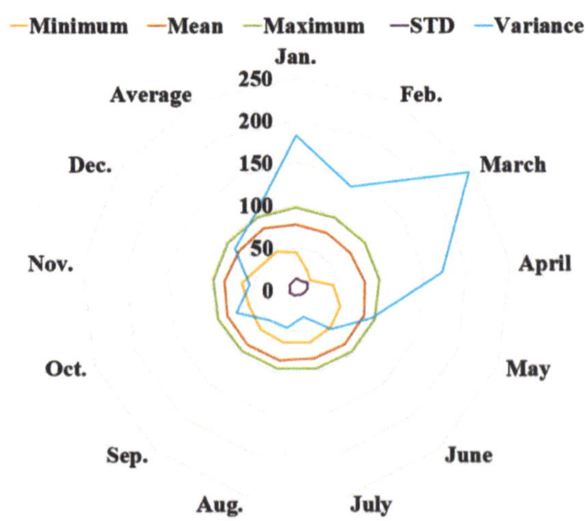

FIGURE 8.8 Relative humidity (%) analysis.

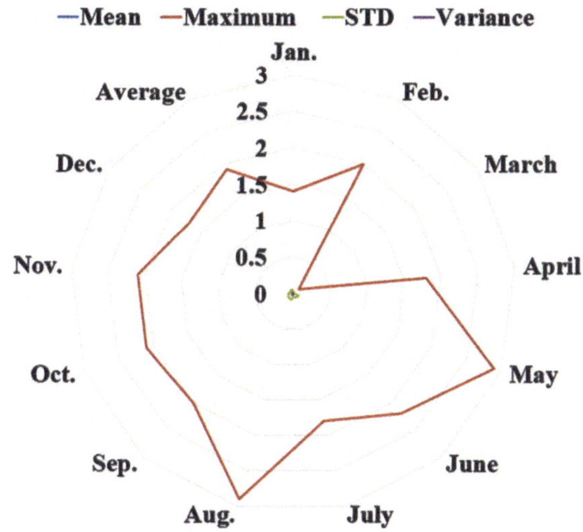

FIGURE 8.9 Precipitation (precip) (in mm).

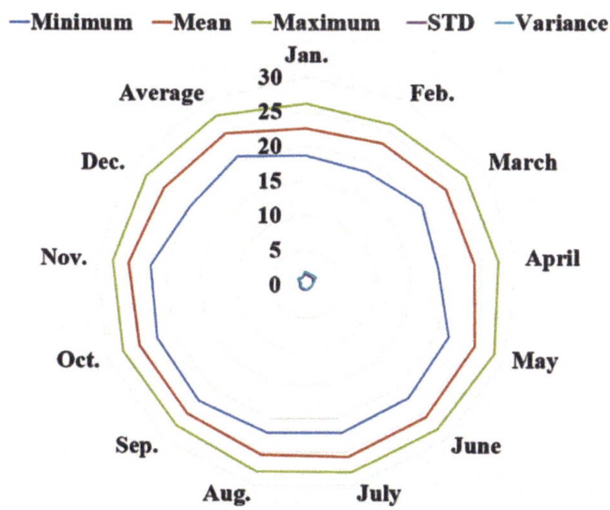

FIGURE 8.10 Wet bulb temperature (wb) (in °C).

the historical data with filling the appropriate missing values and removing the spikes (if any). The purified per month historical data is shown in Fig. 8.11 for better understanding. In Fig. 8.11A–L, the historical and filled missing value data are represented for each month starting from January to December of 2015. The original historical data are represented by dark black color and filled missing value data is shown by gray color for better understanding, and visualization of the missing value pattern obtained by intelligent data analytics.

For further analysis of WSFP using LSTM, the per minute dataset has been used for training and testing phase. The historical data are recorded in 1-min interval. Therefore, total data samples for 1 month are around 44,640 for 1 month and around 5,35,680 in a year, which are divided into two sub-data set (training data~70% and testing data~30%) and testing dataset is divided into multiple files for further use.

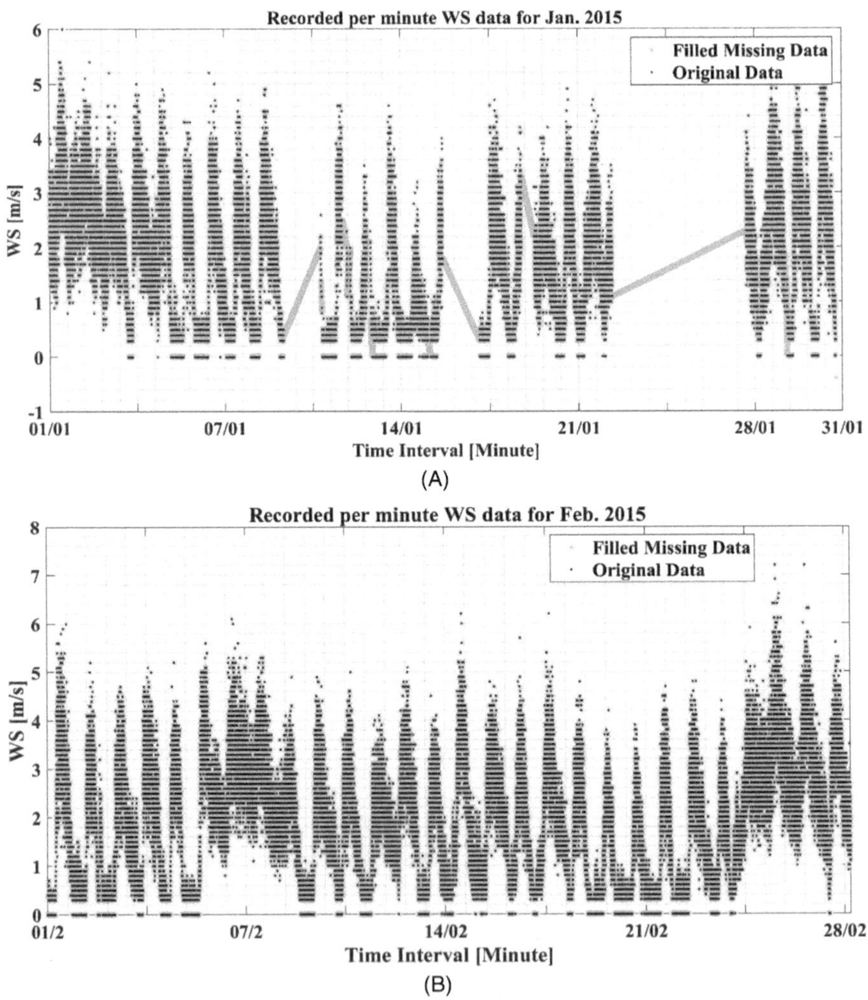

FIGURE 8.11 Per minute WS representation for original historical data and filled missing value data.

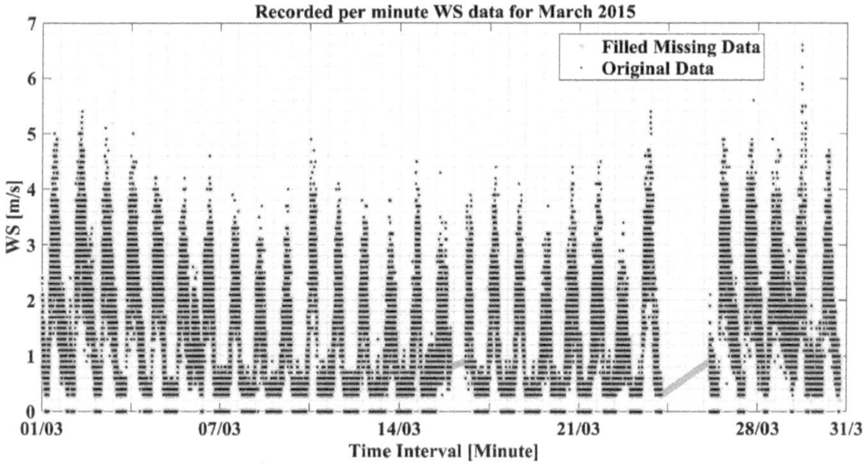

(C)

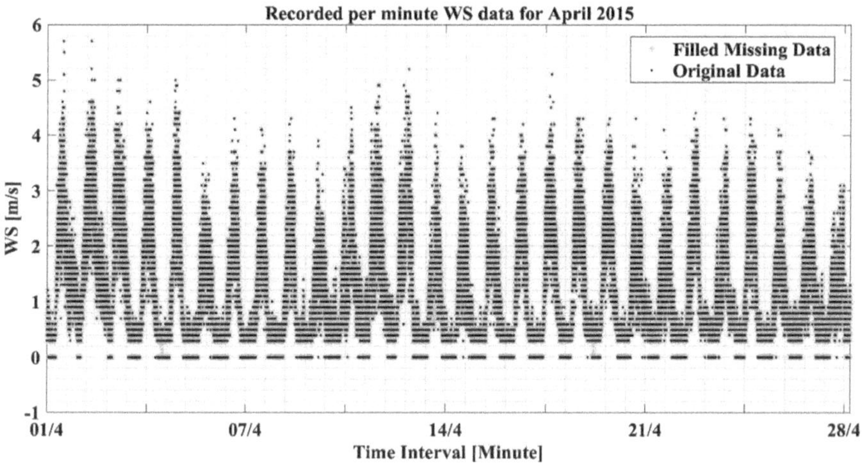

(D)

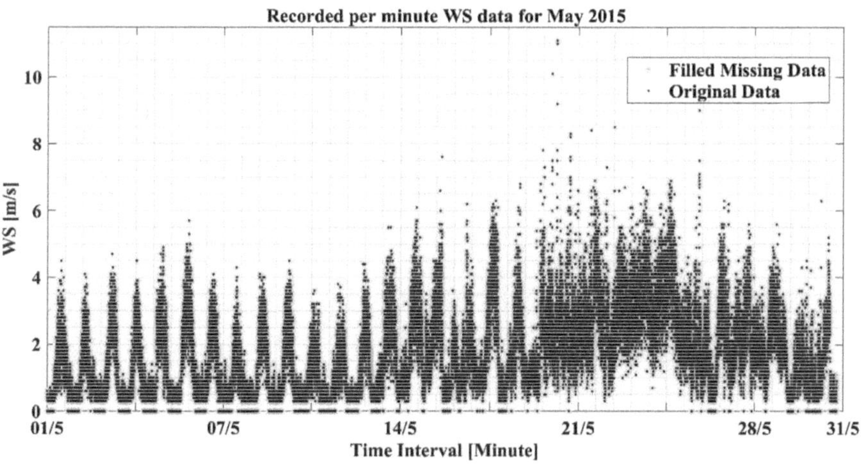

(E)

FIGURE 8.11 (*Cont.*)

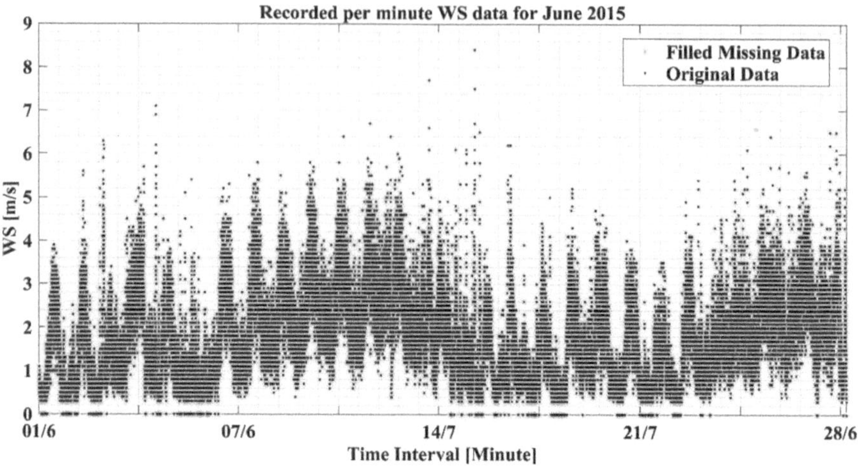

(F)

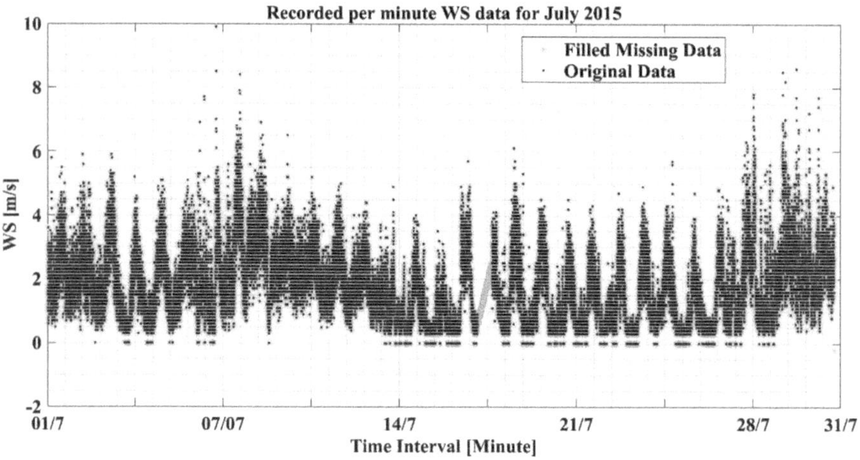

(G)

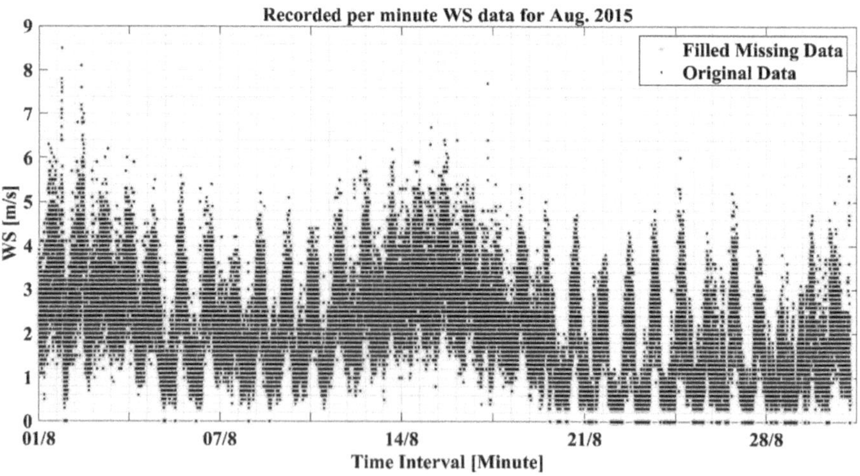

(H)

FIGURE 8.11 (*Cont.*)

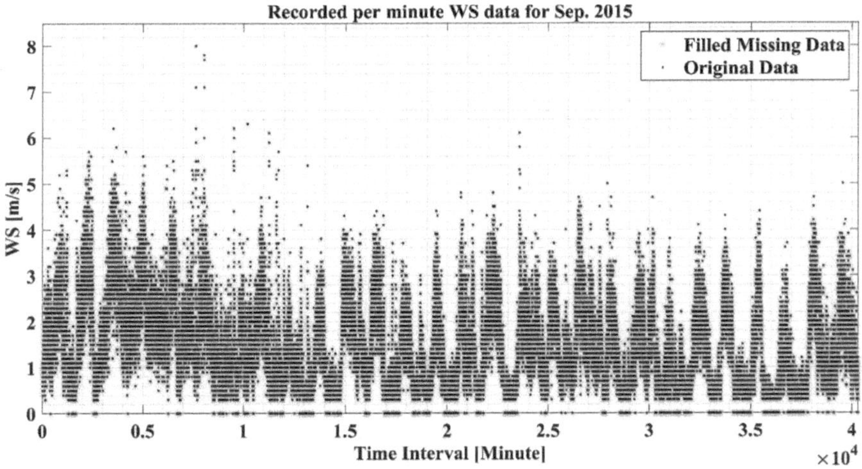

(I)

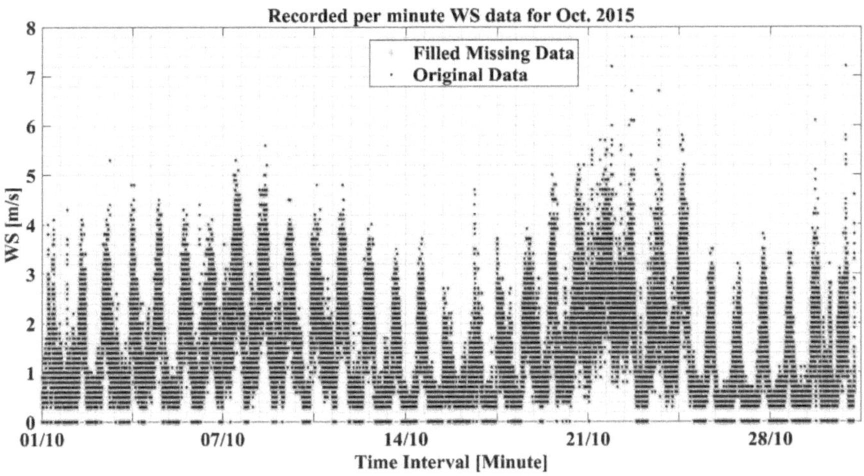

(J)

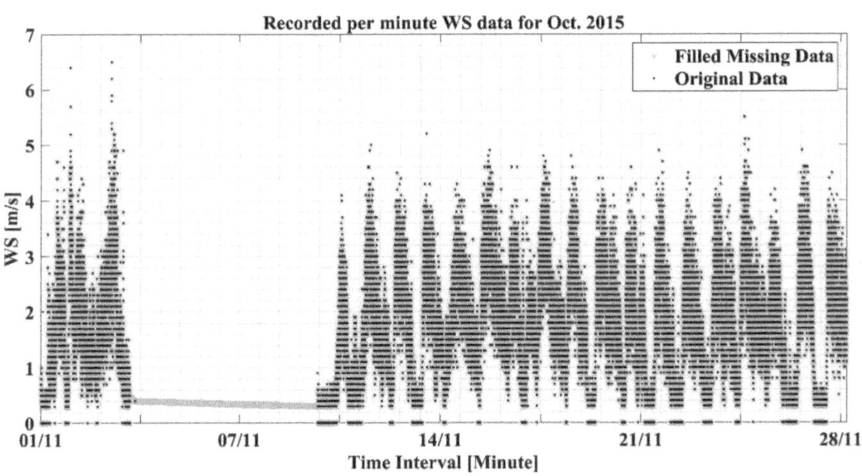

(K)

FIGURE 8.11 (*Cont.*)

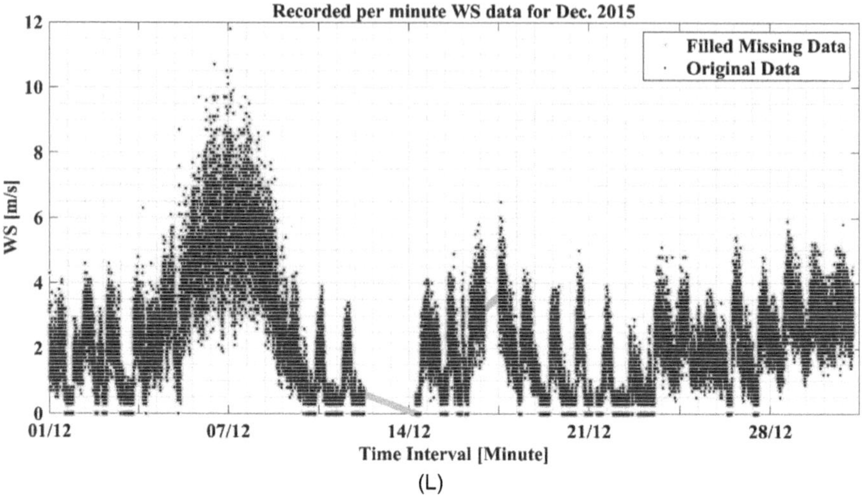

FIGURE 8.11 (*Cont.*)

3.3 Feature extraction

EMD (empirical mode decomposition) method based time-frequency domain features are extracted. EMD extracts the IMFs (intrinsic mode functions) as a number of features that must satisfy 2-basic criteria of the EMD. The implementation of EMD for IMFs extraction is illustrated in Fig. 8.12 [14,15]. These extracted features are utilized as an input variable to LSTM model for WSFP. The length of input feature matrix $\left(IFM_{Att}\right)$ depends on the number of attribute and generated IMFs from raw WS.

Let us assume, there is 1 number of attribute and generated IMFs from each attributes are n then input matrix (IM) is:

$$IFM_{IMFs} = [imf_{1,1}, imf_{1,2}, \ldots.., imf_{1,n}]_{S\times i} \tag{8.1}$$

Where, $i = 1,2,\ldots 3$ (number of attribute/variable), $n = 1,2,\ldots N$ (number of IMFs for each attribute), S = number of data samples.

The generated intrinsic mode functions (IMFs) of 10 numbers for WS are obtained and utilized for further study.

3.4 Most relevant feature selection

Decision tree (DT) based J48 algorithm [16] is used to select the most relevant input variables to the LSTM model. Generally, forecasting accuracy of the LSTM model varies with respect to the number of input variables. So, identification of most relevant variable is performed by J48 algorithm, which is easy

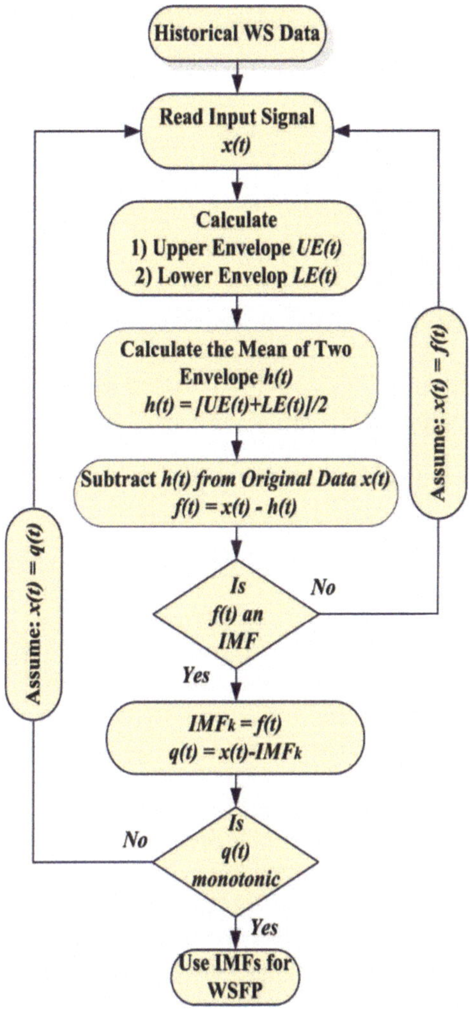

FIGURE 8.12 Empirical mode decomposition (EMD) implementation procedure [14,15].

and very fast in operation and it select only those variables which are most relevant. Least ranked variables are removed from the database.

Utilized the input vector H as $x_i \in R^n,\ i = 1, 2, ..., l$ and target $y \in R^l$, then a DT recursively split space with same target samples are grouped together.

Let data at node m be nominated by Q for each specimen divide $\theta = (j, t_m)$ with feature j and threshold t_m, and data can be divided into $Q_{left}(\theta)$ and $Q_{right}(\theta)$ subgroups as:

$$Q_{left}(\theta) = (x,\ y) \,|\, x_j <= t_m \text{ and } Q_{right}(\theta) = Q \setminus Q_{left}(\theta) \tag{8.2}$$

Impurity at m is evaluated by its function $H()$ according to performed task (such as classification/regression).

$$G(Q,\theta)=\frac{\eta_{left}}{N_m}H(Q_{left}(\theta))+\frac{\eta_{right}}{N_m}H(Q_{right}(\theta)) \tag{8.3}$$

Optimized the parameters to reduce impurity as:

$$\theta^* = \arg\min_\theta G(Q,\theta) \tag{8.4}$$

Iterate again-and-again for $Q_{left}(\theta^*)$ and $Q_{right}(\theta^*)$ till $N_m < \min_{samples}$ or $N_m = 1$.

If problem is formulated for classification, then target is $0,1,....,K-1$, for m node, and notifying a region R_m with N_m instances is $p_{mk} = 1/N_m \sum_{x_i \in R_m} I(y_i = k)$.

Generally, evaluation of impurity (i.e., Gini):

$$H(X_m)=\sum_k p_{mk}(1-p_{mk}) \tag{8.5}$$

$$\text{Cross entropy}: H(X_m)=-\sum_k p_{mk}\log(p_{mk}) \tag{8.6}$$

$$\text{Misc Identification}: H(X_m)=1-\max(p_{mk}) \tag{8.7}$$

After implementation of the J48 algorithm for most relevant IMFs selection, it is found that pruning of the number procedure for WSFP problem is not effective and it does not enhance too much forecasting accuracy. So, key take away message is that feature selection for forecasting the WS may not enhance the prediction accuracy at very high level as compared to feature selection based method. So to reduce the computational burden, it may not be opted for the real-site implementation.

3.5 Design of LSTM network

Conventional neural network (NN) is restricted to flow of sequence data and also restricted to parameter sharing. To overcome this problem, RNN was introduced, which has a hidden state and less memory unit to store the parameter of previous state. According to available information, RNN may handle long sequence. However, in actual condition, networks could not handle long term sequences [17] and suffers with either of vanishing gradient [18] or exploding gradient problems [19]. To overcome these limitations, special variant of RNN called long short-term memory (LSTMs), was introduced [20]. Similar to RNNs, LSTMs also have the sequential structure with four states. Here in this chapter, the architecture of LSTM has been modified (as shown in Fig. 8.13) in

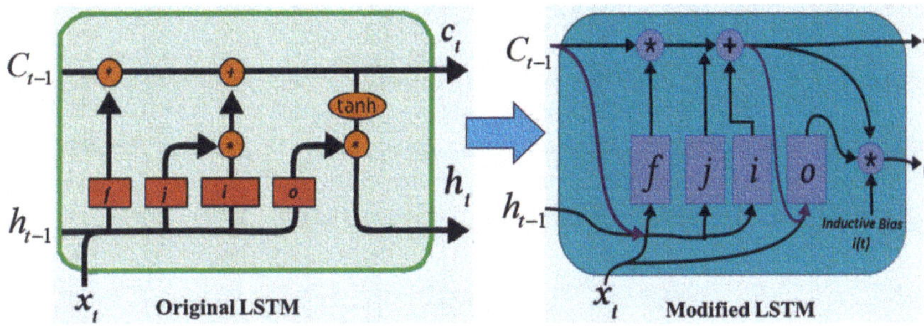

FIGURE 8.13 LSTM and modified-LSTM cell representation.

order to adapt with missing cases and inductive bias term is added to compensate the missing features.

The fundamental component of LSTMs is a cell state, it acts like the assembly line of the network by carrying information between different cells. Input gate adds the information from the previous hidden state vector with current input and the resultant values are passed through *Tanh* activation function. Forget gate with sigmoid activation function decides which part of the information to retain by passing previous hidden state and current input through sigmoid activation function. Values closer to 1 are retained and closer to 0 are discarded. Output gate decides the hidden state for next cell state. Equations to calculate the different cell states are given as:

$$Sigmoid = 1/(1+e^{(-g)}) \tag{8.8}$$

where g in sigmoid function corresponds to function with respect to below gates.

$$forget\ gate = \sigma\left(W_f\left[S_{\{t\text{-}1\}}, X_t\right] + K_f\right) \tag{8.9}$$

$$Input\ gate = \sigma\left(W_i\left[S_{\{t\text{-}1\}}, X_t\right] + K_i\right) \tag{8.10}$$

$$Output\ gate = \sigma\left(W_o\left[S_{\{t\text{-}1\}}, X_t\right] + K_o\right) \tag{8.11}$$

$$h(t) = \left(\left(forget * C_{\{t\text{-}1\}}\right) + \left(input * C_t\right)\right)\ i(t) \tag{8.12}$$

Where, W_f, W_i, and W_o are called weights of forget, input, and output gates, respectively. K_f, K_i, and K_o corresponds to bias terms for forget, input, and output gates. $S_{\{t\text{-}1\}}$ corresponds to previous hidden state vector and X_t is the input for time step t. In the final equation, $h(t)$ is the hidden state vector of current cell.

3.6 Performance measure indices

The performance evaluation process is used to compute the model's characteristics in form of its performance, which shows the model's capability to STFP. In this study, MAE (mean absolute error), MAPE (mean absolute percentage error), and RMSE (root mean square error) are computed. These performance indices computed for the DNN model are given as [21]:

$$MAE = \frac{1}{N}\sum_{t=1}^{K} \left| \hat{f}(t) - f_h(t) \right| \quad (8.13)$$

$$MAPE = \left(\frac{1}{N}\sum_{t=1}^{K} \left| \frac{\hat{f}(t) - f_h(t)}{f_h(t)} \right| \right) \times 100 \quad (8.14)$$

$$RMSE = \sqrt{\frac{1}{N}\sum_{t=1}^{K} \left(\hat{f}(t) - f_h(t) \right)^2} \quad (8.15)$$

Where, $\hat{f}(t)$ and $f_h(t)$ are the forecasted and historical solar radiation at time interval k respectively, K is the total number of forecasted and historical data samples.

4 Case study: demonstration of results and discussion

This section represents the WS forecasting obtained results with the help of proposed LSTM and modified-LSTM (mLSTM) for different time intervals of one-step-ahead and forecasted upto step four. For each step ahead forecasting of WS, the performance indices (MAE, MAPE, and RMSE) are evaluated for each LSTM and mLSTM model, which is shown in Table 8.9 for the one-step ahead forecasting. The Table 8.9 shows the outperformed results of proposed LSTM and mLSTM models. Then based on these performance evaluation, mLSTM model was utilized to forecast the WS with two-step ahead, three-step ahead, and four-step ahead with the same dataset with different time interval. In the multi-step forecasting, following assumptions were considered: one-step (1 min), two-step (2 min), three-step (3 min), and four-step (4 min) ahead forecasting. The learning cure for training phase of mLSTM model is represented in Fig. 8.14, in which dark blue and dark orange color lines represented the training (smoothed) accuracy and loss function of the model, respectively. Similarly, light blue and light orange color lines have represented the training (output) accuracy and loss function of the model respectively, which are very negligible in this case.

TABLE 8.9 IMFs based one-step ahead WS forecasting performance indices representation.

		MAE		RMSE		MAPE (%)	
		LSTM	mLSTM	LSTM	mLSTM	LSTM	mLSTM
2015 Dataset	Training phase	0.451	0.152	0.734	0.425	2.35	1.984
	Testing phase	0.825	0.484	1.051	0.852	6.21	5.945
2016 Dataset	Testing phase	0.985	0.765	3.235	2.540	8.24	9.50
2017 Dataset	Testing phase	0.992	0.756	3.05	1.985	9.8	9.885

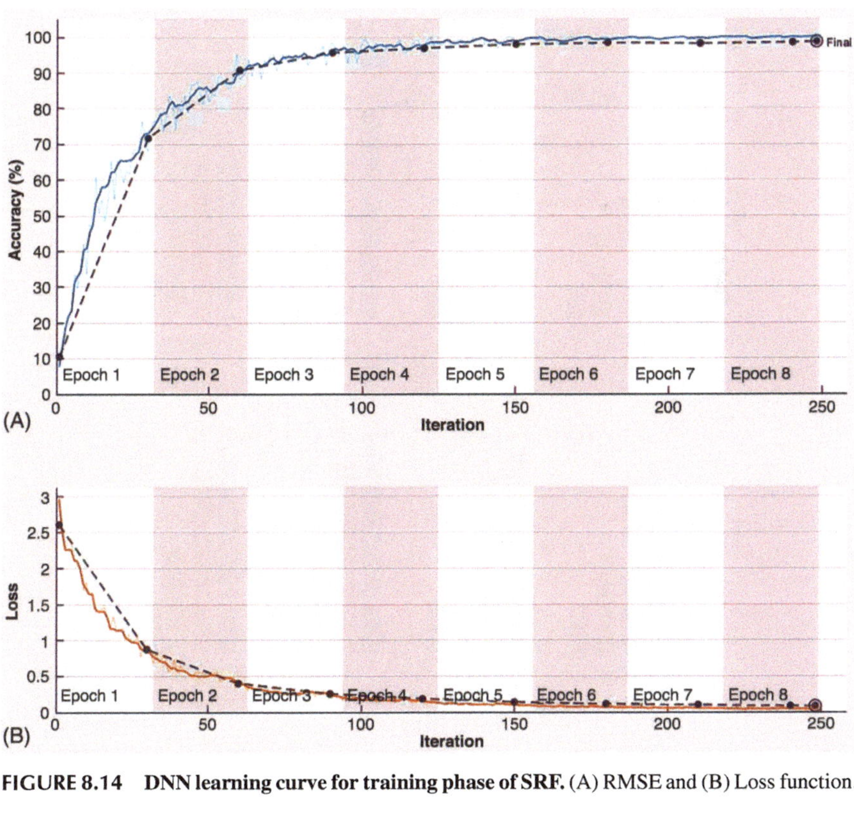

FIGURE 8.14 DNN learning curve for training phase of SRF. (A) RMSE and (B) Loss function.

The Fig. 8.15 represents the forecasting behavior and performance indices obtained with the proposed mLSTM for overall time period of records. Presented results show that proposed mLSTM method outperforms for all cases of multi-step, which presents the acceptable value of MAE, RMSE, and MAPE error. The capability of the proposed mLSTM method is demonstrated for its effectiveness by using Port Blair historical dataset from 2015 to 2017 and the testing phase results for each month, which shows the high level of acceptability of the proposed mLSTM model.

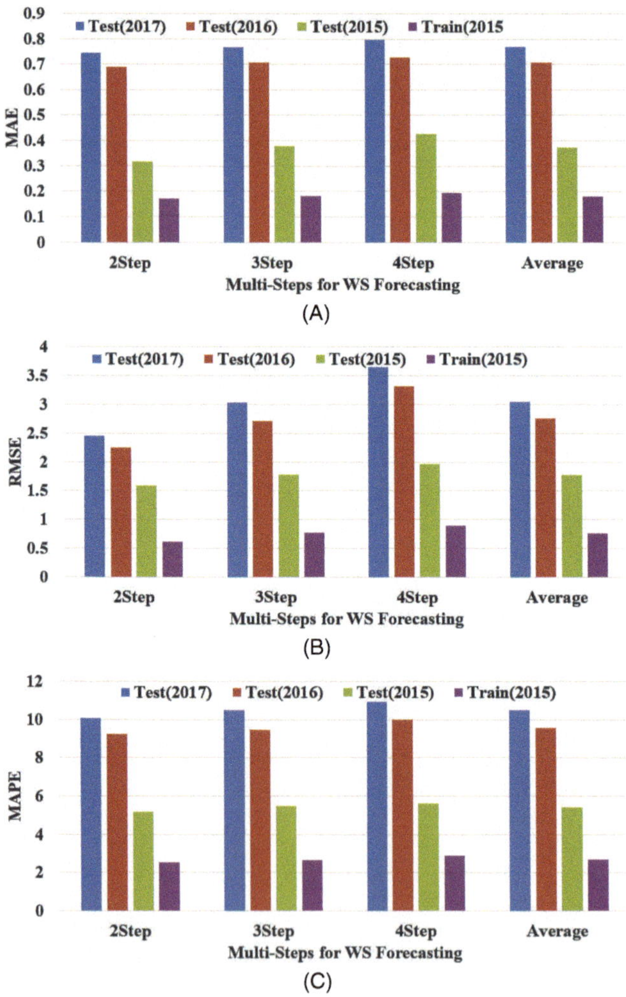

FIGURE 8.15 The representation of WSF model performance indices for multi-step forecasting. (A) MAE, (B) RMSE, and (C) MAPE.

5 Conclusion

A new advanced mLSTM approach is proposed for multi-step ahead WSF and its performance is validated and demonstrated by using Indian city of Port Blair for short term forecasting of WS. This network model is capable of solving the linear as well as non-linear problems. So, WSF is a highly non-linear problem which is short term forecast (per minute), which is handled by proposed mLSTM method in an efficient way with less computational burden. The multi-step ahead forecasting is performed recursively, using historical datasets one-step (1 min), two-step (2 min), three-step (3 min), and four-step (4-min) ahead forecasting, and this procedure is repeated until a time step of 4 min ahead. The approach is demonstrated using real meteorological data from Port Blair, a city of India. The demonstration of the proposed mLSTM model is accomplished comparing the obtained results with LSTM model. The demonstrated results showed that the proposed mLSTM model for WSF outperformed with respect to the all criteria (MAPE, RMSE, and MAE) of ANN model for one-step ahead forecasting results.

References

[1] Renewable energy systems: Technology overviews and perspectives, in: F. Blaabjerg, D.M. Lonel (Eds.), Renewable Energy Devices and Systems with Simulations in MATLAB and ANSYS, CRC Press LLC, 2020.

[2] H. Ritchie, M. Roser, Renewable Energy. Available from: https://ourworldindata.org/renewable-energy.

[3] Ministry of New and Renewable Energy Resources (MNRE). Available from: https://mnre.gov.in/file manager/annual-report/2016-2017/EN/pdf/1.pdf.

[4] Ministry of New and Renewable Energy Resources (MNRE). Available from: https://mnre.gov.in/file-manager/akshay-urja/october-2017/Images/20-25.pdf. Accessed 22.02.2019.

[5] D.B. Alencar, et al., Hybrid approach combining SARIMA and neural networks for multi-step ahead wind speed forecasting in Brazil, IEEE Access 6 (2018) 55986–55994, doi: 10.1109/ACCESS.2018.2872720.

[6] K. Gnana Sheela, Neural network based hybrid computing model for wind speed prediction, Neurocomputing 122 (2013) 425–429, http://dx.doi.org/10.1016/j.neucom.2013.06.008.

[7] L. Wang, X. Lid, Y. Bai, Short-term wind speed prediction using an extreme learning machine model with error correction, Energy Convers. Manage. 162 (2018) 239–250, https://doi.org/10.1016/j.enconman.2018.02.015.

[8] Z. Su, J. Wang, H. Lu, G. Zhao, A new hybrid model optimized by an intelligent optimization algorithm for wind speed forecasting, Energy Convers. Manage. 85 (2014) 443–452, http://dx.doi.org/10.1016/j.enconman.2014.05.058.

[9] J. Shi, J. Guo, S. Zheng, Evaluation of hybrid forecasting approaches for wind speed and power generation time series, Renew. Sustain. Energy Rev. 16 (2012) 3471–3480, doi: 10.1016/j.rser.2012.02.044.

[10] E. Cadenas, W. Rivera, Wind speed forecasting in three different regions of Mexico, using a hybrid ARIMA-ANN model, Renew. Energy 35 (2010) 2732–2738, https://doi.org/10.1016/j.renene.2010.04.022.

[11] M. Yesilbudak, S. Sairoglu, I. Colak, A new approach to very short term wind speed prediction using k-nearest neighbor classification, Energy Convers. Manage. 69 (2013) 77–86, https://doi.org/10.1016/j.enconman.2013.01.033.

[12] J. Wang, Y. Songa, F. Liu, R. Hou, Analysis and application of forecasting models in wind power integration: A review of multi-step-ahead wind speed forecasting models, Renew. Sustain. Energy Rev. 60 (2016) 960–981, https://doi.org/10.1016/j.rser.2016.01.114.

[13] A.K. Yadav, H. Malik, S.S. Chandel, Selection of most relevant input parameters using WEKA for artificial neural network based solar radiation prediction models, Renew. Sustain. Energy Rev. 31 (2014) 509–519, doi: https://doi.org/10.1016/j.rser.2013.12.008.

[14] H. Malik, S. Mishra, Artificial neural network and empirical mode decomposition based imbalance fault diagnosis of wind turbine using TurbSim, FAST and Simulink, IET Renew. Power Generat. 11 (6) (2017) 889–902, doi: 10.1049/iet-rpg.2015.0382.

[15] H. Malik, S. Mishra, Application of fuzzy Q learning (FQL) technique to wind turbine imbalance fault identification using generator current signals, in: Proceedings of IEEE Seventh Power India International Conference, Bikaner, India, 2016, pp. 1–6, doi: 10.1109/POWERI.2016.8077283.

[16] H. Malik, S. Mishra, A.P. Mittal, Selection of most relevant input parameters using Waikato environment for knowledge analysis for gene expression programming based power transformer fault diagnosis, Int. J. Electric Power Compon. Syst. 42 (16) (2014) 1849–1862, doi: http://dx.doi.org/10.1080/15325008.2014.956952.

[17] Y. Bengio, P. Simard, P. Frasconi, Learning long-term dependencies with gradient descent is difficult, IEEE Trans. Neural Networks 5 (2) (1994) 157–166.

[18] S. Hochreiter, The vanishing gradient problem during learning recurrent neural nets and problem solutions, Int. J. Uncertain. Fuzziness Knowledge Syst. 6 (02) (1998) 107–116.

[19] R. Pascanu, T. Mikolov, Y. Bengio, Understanding the exploding gradient problem, ArXiv abs/1211.5063 2 (2012) 417.

[20] S. Hochreiter, J. Schmidhuber, Long short-term memory, Neural Comput. 9 (8) (1997) 1735–1780.

[21] A.K. Yadav, V. Sharma, H. Malik, S.S. Chandel, Daily array yield prediction of grid-interactive photovoltaic plant using relief attribute evaluator based radial basis function neural network, Renew. Sustain. Energy Rev. 81 (2) (2018) 2115–2127, doi: https://doi.org/10.1016/j.rser.2017.06.023.

Chapter 9

Intelligent Data Analytics for Time-Series Load Forecasting Using Fuzzy Reinforcement Learning (FRL)

1 Introduction

The electric load forecasting (ELF) is indispensable procedure for the planning of power system industry, which plays an essential role in the scheduling of electricity and the management of the power system (PSM). Hence, ELF in advance stage has numerous great values for managing the generation capacity, scheduling, management, peak reduction, market evaluation, etc. Therefore, the forecasting as per the time-horizon has extremely useful for its advantages to fulfill the several requirements as per their application as shown in Table 9.1.

Generally, load forecasting (LF) refers to the expected load requirements, which is computed by utilizing a systematic procedure of defining future prospective loads in sufficient quantitative information, which is used for taking decision for system expansion. Moreover, LF required for the following applications such as: (1) capacity planning, (2) network planning, (3) generation & transmission capital investment planning, (4) financial forecast, (5) efficient power procurement, (6) selling of excess power, (7) fuel ordering planning, (8) optimal supply and scheduling, (9) renewable planning, and (10) fuel mix selection planning, etc. In the power system operation and planning (PSOP), the LF is categorized into three groups as: (1) long-term forecasting (LTF): 1–20 years, (2) medium-term forecasting (MTF): 1 week–1 year, and (3) short-term forecasting (STF): 1 h–1 week. The LTF plays a fundamental role in economic planning of new future coming generation and its transmission. The MTF is mainly useful for the scheduling of supply fuel, maintenance planning, financial planning, and formulation of tariff. And the STF is used to provide the fundamental information for planning start-up and shut-down scheduling of the generating unit, spinning reserve planning, and detailed study of transmission constraints. The STF is also utilized for ELD and system security assessment. Here, one important take away massage for the reader is that the forecasting

Intelligent Data-Analytics for Condition Monitoring. http://dx.doi.org/10.1016/B978-0-323-85510-5.00009-0

TABLE 9.1 Time-horizon based application area of the forecasting mechanisms.

Time-horizon perspective	Application area	Remarks
1–10 years	Power system planning (PSP)	It includes generation planning, network planning and load forecast. The basic structure of PSP is shown in Fig. 9.1.
1 week to 1 year	Maintenance scheduling	
1 min to 1 week	Unit commitment analysis	
	Economic load dispatch (ELD) and OPF analysis	
	Automatic generation control	
Milliseconds to seconds	Power system dynamic analysis	
Nanoseconds to micro seconds	Power system transient analysis	

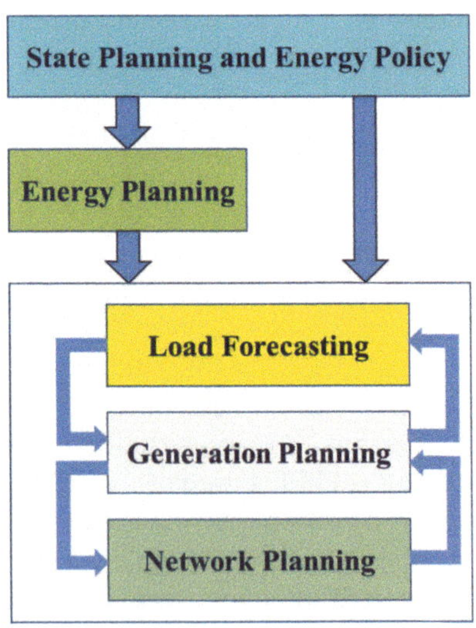

FIGURE 9.1 Basic structure of PSP.

accuracy decreases with respect to the time-horizon of the forecast, which plays an important role for all applications in this domain.

There are several advantages of the LF, which are: (1) LF ensures the availability of supply of electricity, (2) LF provides the means of avoiding over &

underutilization of generating capacity, (3) LF helps to make use of best possible use of capacity, (4) LF save unnecessary capital expenditure, and (5) LF prevents optimum economic growth, etc., in spite of some advantages, there are some uncertainties of LF. These uncertainties are (1) forecasting the future requirements/needs of electricity, (2) electricity production & distribution are highly capital intensive, (3) projects lead times are very long due to large size of project, (4) forecasting is not an isolated activity, (5) electrical energy role should be highlighted in the society, (5) the national policy & strategy are very key-points, (6) it may affect due to policies, public perception, & viewpoints, (7) additional needs for LF is DSM (demand side management) and conservation policies, (8) to make highly preciseness, (9) single LF can be risky, (10) always be ready to change as per the situation. Therefore, the LF depends on several factors, which affect the forecast performance. These affecting factors are: use of land, city plans, industrial plans, community development plans, alternative energy resources, load density, population growth, historical data, and geographical factors. So, in the LTF of the load structure, two different categories of data (historical data & weather data and forecasted exogenous variables) are used for LF results for upto 20 years. And the key modeling techniques used for LTF are: trend analysis, linear multiple variable regression, partial end use method, and several scenario approach. Similarly, in STF of the load structure, three different categories of data (historical load & weather data, real-time load data from SCADA/sensor trough energy meters, and forecasted exogenous variables) are used for LF results for up to next 24 h or week. The key techniques are used for STF are: time-series analysis (AR, ARX, ARMA, etc.), multiple linear regression, advanced approaches (i.e., neural networks, fuzzy logic, SVM, etc., and its hybrid mode). Therefore, widely used electric LF models that have been published by the several researches in the digital domain are summarized in Fig. 9.2 of Section 3. In recent years, several advanced

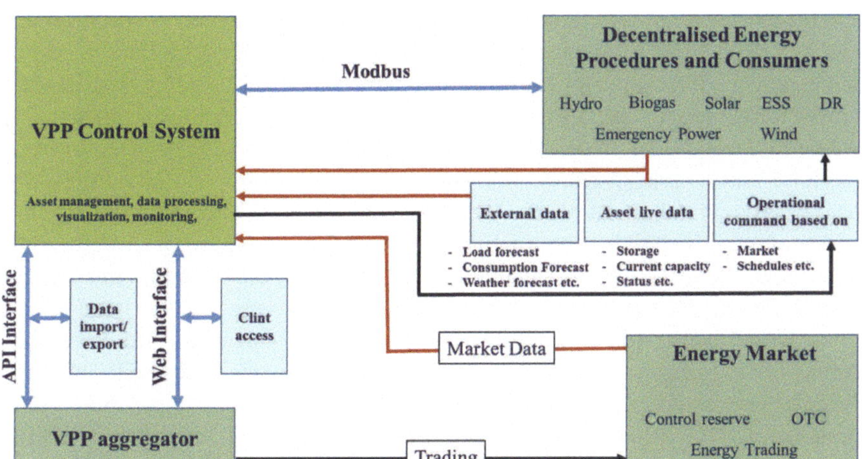

FIGURE 9.2 Representation of a high level architecture of VPP.

approaches have been implemented in this area to overcome some drawbacks of the other technique [1–5]. But most of approaches required system parameters, which are not possible to fetch out in an easy way. Therefore in this chapter, fuzzy-Q-learning (FQL) based fuzzy reinforcement learning (FRL) approach is implemented to solve this problem. The main beauty of proposed FRL approach is that it gives optimal results without observing system parameters. Moreover, proposed approach is not data dependent. It harnesses the information from the property of historical dataset only.

The presented chapter is organized into five sections. Section 1 represents the introduction part of the study, which represents the need of ELF. Section 2 represents the intelligent data analytics for LF which includes the market value and inventions details of the LF research area. In Section 3, methodology is presented, which includes the proposed approach, brief details of FRL approach, and used dataset for the demonstration. In Section 4, four different case studies of LF are presented to demonstrate the performance ability of the proposed approach. Finally, conclusion of the work and future scope is presented in Section 5.

2 Intelligent data analytics for load forecasting

In this section two types of analysis have been presented. The first analysis is related to the market growth and requirement and interlink of the LF in the core area of the power system with respect to the market demand and future prospective research. The second analysis is related to the inventions which have been invented in the area of the ELF. This analysis is very useful to identify the green zone area for the future coming research for load forecasting and its associated sub- areas/regions.

In the modern world, LF is a key factor/attribute, which is used to manage the power plant. In the recent scenario, a new term named "Virtual Power Plant (VPP)" came in picture, which is the network of decentralized, medium scale power generating units (i.e., solar park, wind farm, combined heat & power unit, and storage system, etc.). The interconnected units are dispatched via central control room of VPP. The main objective of VPP is to share/relieve the load on the grid in smart way distribution to manage and smoothing the peak-load demand. Most of the advanced work related to VPP is from Europe region, which has high value of future market around ~\$470 million in 2030 with expected CAGR (compound annual growth rate) of ~18%plus in comparison to 2019 (~\$75 million). Initially, ESS (energy storage system) and DERs (distributed energy resources) were the part of VPP but now more than four sub-systems (DERs, ESS, DSM-demand side management, and EVs-electric vehicles) are the part of the VPP, which makes more reliable. The main components of the VPP are hardware (i.e., VPP control box), software (i.e., VPP central system, software maintenance and support), and its services (i.e., DR related software implementation, energy trading). A high-level architecture of VPP is shown in Fig. 9.2. According to the hardware point of view, VPP system's main components are: (1) DERs, (2) ESS, (3) controllable load, and (4) ICT (information communication technologies). The DERs has the power generation sub-systems and utilities (RES: PV, wind, hydro, etc., and non-RES: CHP, biomass, biogas, diesel, gas, fuel cell, etc.). Generally, power generation capability for different areas are: (1) for residential: 1–10 kW, (2) for industrial: >300 kW, and (3) for commercial: 50–300 kW. Here in VPP, ESS can be utilized for on-peak period and may store on off-peak period. Generally, ESS capacity are: (1) for residential: <5 kVA, (2) for industrial: >20 kVA, and (3) for commercial: 5–20 kVA. The third component of VPP is controllable loads, which is referred to DR analysis. Finally, ICT is responsible to connect the all sub-systems and its associated equipment to form a VPP so that following work can be performed in a systematic way: (1) monitoring the status of GU, (2) forecasting demand, load, supply, (3) to provide the coordination between elements of VPP, (4) energy trading, and DERs, ESS, loads operation control, etc.

A VPP is an advanced technology which is scalable from a small community to a national level. So, as per geographical size, VPP can be categorized in to three main types as given in Table 9.2 as follows: (1) local LVPP or community based (CVPP), (2) regional based (RVPP), and large scale based (LSVPP). The current status of VPP technologies is at LVPP (type#3). Therefore, in the market, there are different VPP business models which are used by the policy makers. These business models are: (1) supply side VPP: it includes DER, (2) demands side VPP: it includes batteries, generators, and HVAC (heating, ventilation, and air conditioning) system, and (3) mixed asset APP: it includes both type of VPP. There are some companies which provide the services for the VPP as: for supply side VPP: ABB, Siemens, Next Kraftwerke, VPP emergy, LichtBlick, Statkraft, etc. For demands side VPP, companies are: AutoGride, Restore, KiWi Power, Next Kraftwerke. And finally, for mixed mode VPP, companies are: ABB, Siemens, Next Kraftwerke.

In the area of invention and innovative research for the ELF, data starting from 2001 to 19 has been analyzed. In the analysis, percentage value of yearly applied total applications versus number of granted application has been represented in Fig. 9.3, which shows that the granting rate is going

TABLE 9.2 VPP classifications according to the geographical size.

Type#1	LSVPP or CVPP				In 5–10 years
Type#2	RVPP1	RVPP2	RVPPn	System elements	In 3–5 years
Type#3	LVPP1	LVPP2	LVPPn	System elements	Current status
	Trade system	Balancing system	Network support system	System elements	

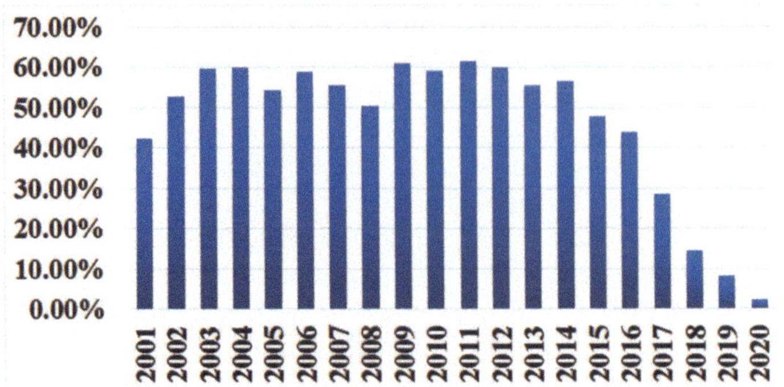

FIGURE 9.3 **Per year innovation representation.**

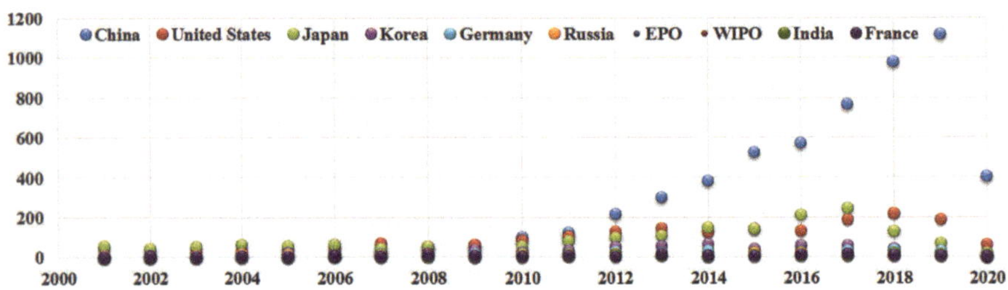

FIGURE 9.4 **Representation of invention applications applied by the top 10 countries.**

down year-wise-year stared from 2015. Generally, three types of innovations are available in ELF area (i.e., invention: ~95% plus, utility: ~4% plus, and design ~0.01% plus). The top 10 counties for these innovations are as per the rank: China, Japan, United States, Korea, Germany, Russia, EPO (European Patent Office), France, WIPO (World Intellectual Property Office), and India. The invention applications applied by the top 10 countries per year is shown in Fig. 9.4. The application trend shows a significant sign that numbers of inventions are increasing and China is at the top.

3 Time-series load forecasting model

In this section, a detailed analysis of different type of load forecasting models have been studied and categorized according to the application and innovation part of the model as represented in Fig. 9.5. Generally, time-series load forecasting models are classified into three categories, which are: (1) statistical models, (2) artificial intelligent (AI)/machine learning (ML) models, and (3) hybrid models. The statistical models are generally mathematical model which are based on a set of statistical assumption and hypothesis. This type of model represents the mathematical relationship between one/more random attributes and other non-random attributes. Several statistical models have been developed in the area of forecasting, some of them are listed as follow: (1) AR (autoregressive)

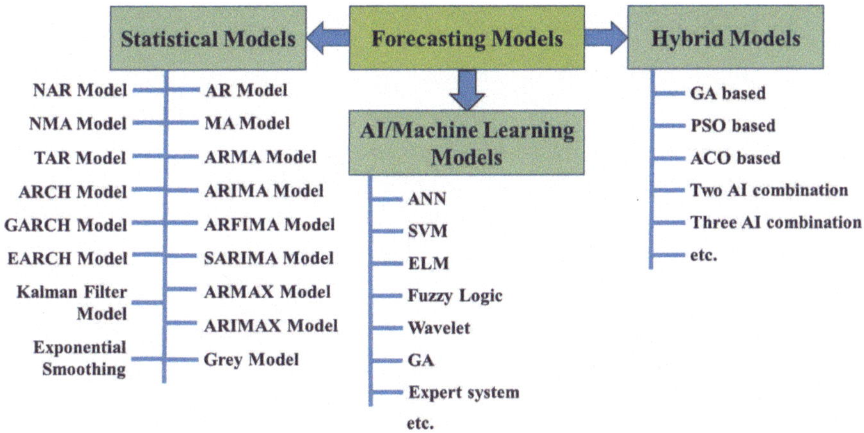

FIGURE 9.5 **Forecasting models for load forecasting.**

model, (2) MA (moving average) model, (3) ARMA (autoregressive moving average) model, (4) ARIMA (autoregressive integrated moving average) model, (5) ARFIMA (autoregressive fractionally integrated moving average) model, (6) SARIMA (seasonal autoregressive integrated moving average) model, (7) ARMAX model, (8) ARIMAX model, (9) NAR (nonlinear autoregressive) model, (10) NMA (nonlinear moving average) model, (11) TAR (threshold autoregressive) model, (12) ARCH (autoregressive conditional heteroskedasticity) model, (13) GARCH (generalized ARCH) model, (14) EARCH (exponential generalized ARCH) model, (15) Kalman filter model, (17) ES: exponential smoothing, and (18) GM: Grey model. These statistical models have been further categorized into two main groups: linear time series models and Nonlinear time series model.

Apart of statistical model, there are several models of AI/machine learning and its combination mechanism (hybrid models). Generally, conventional statistical models have some limitations and lead unsatisfactory performance for some problem. Moreover, convention methods are prone to high computational burden, which lead high processing time duration. Therefore, due to these reasons and property of nonlinear dataset, AI/ML based model come in picture in this research domain. There are numerous AI/ML approaches such as: (1) ANN (artificial neural network) model, (2) SVM (support vector machine) model, (3) ELM (extreme learning machine) model, (4) fuzzy-logic model, (5) wavelet based model, (6) GA (genetic algorithm) model, (7) expert system, (8) GEP (gene expression programming) model, etc. Numerous applications are available in the digital domain by using these AI/ML models in different field too. The hybrid models are generally developed by combination/unification of two or more than two AI/ML algorithms. Some examples are: GA based, PSO (particle swarm optimization) based, ACO (ant colony optimization) based, GP (genetic programming based) based, etc.

4 Methodology

4.1 Proposed approach

Fig. 9.6 represents the proposed approach for LF using FQL based FRL approach, which is developed into four (4) parts such as: (1) data collection and transmission, 2) data processing part, 3) FRL model development, and training part, and (4) model testing and adaptability checking. First two parts include

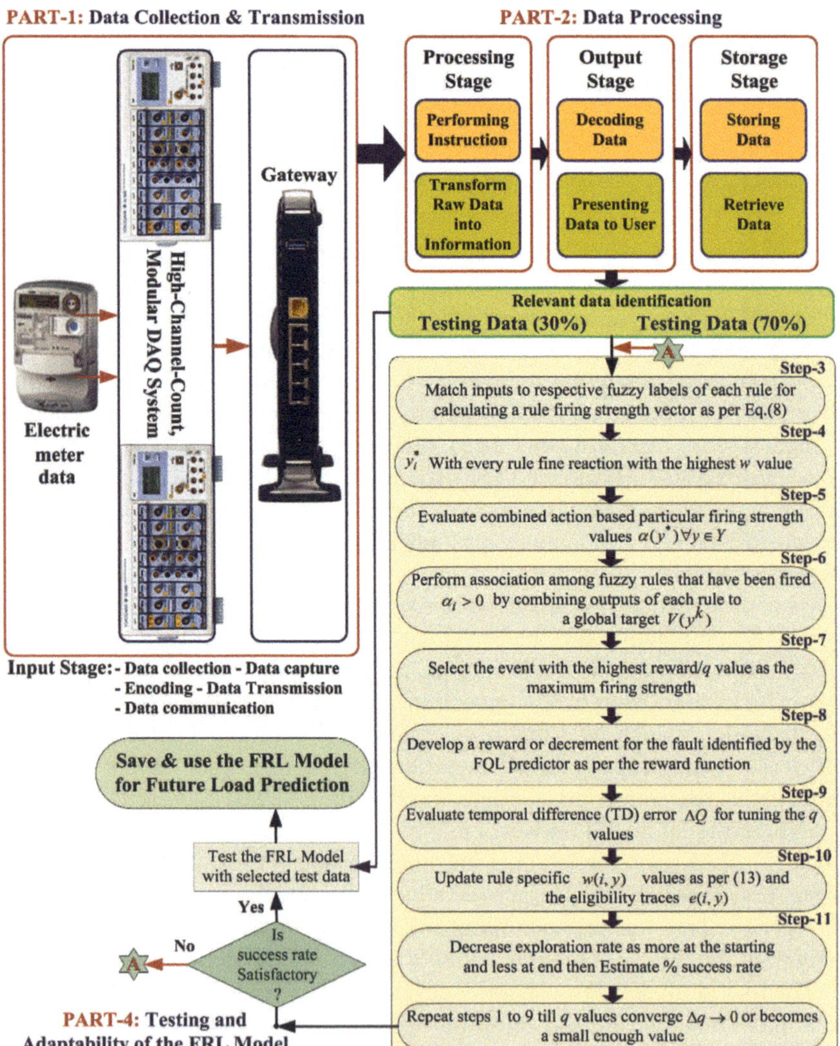

FIGURE 9.6 Proposed hybrid model for load forecast.

11 basic steps. Part-1 comprises three steps (i.e., electric meter connection DAQ for data collection then data transmission through gateway) and Part-2 includes four steps (i.e., processing stage, output stage, data storage stage, and relevant data identification). The dataset collection, data preprocessing (spikes removal and missing value filling), most relevant variable selection, data reduction, and six different data files (i.e., training, testing files for hourly, weekly, and monthly forecasting) formation are included. The training data file is used for FRL model development and developed FRL model learns the ELF property based on historical dataset. After training phase, testing phase is performed to test the trained model performance. If testing phase performance is acceptable then further forecasting for unlabeled data file is proceed. Therefore, trained model is tested in multiple times to cross check the model performance by using different datasets. The ELF model using fuzzy RL is designed and validated with 2004–08 historical dataset of GEFcom2012 [6]. The main advantages of the proposed approaches are:

- Real-time recoded data-based study is possible
- Data preprocessing remove the spikes and fill the missing values, if any, which makes the model more reliable
- Self-adaptability nature of the model makes approach more robust for real-time implementation
- Multiple step testing defines the model acceptability limit in global manner

4.2 Brief detail of FRL approach

The implementation of FRL based FQL based LFM (load forecasting model) has been presented here along with parameter value is given as follows and summarized into 11 steps. For more detail, reader may refer to Refs. [6–10].

Step#1: Match Input *input's* with particular labels and generates rules for rule firing strength vector α_i $\forall i \in N$ as per Eq. (9.1):

$$\alpha_{input}(c_1) = e^{-(c_1 - j_1)^2/2\sigma_1^2} \tag{9.1}$$

Where, j_1 and σ_1 is the central value and standard deviation = $(c_{\max} - c_{\min})/5$.

Step#2: With every rule fine reaction with the highest w value y_i^*.
Step#3: Evaluate combined action based particular firing strength values $\alpha(y^*)\ \forall y \in Y$.
Step#4: Perform association among fuzzy rules that have been fired, that is, $\alpha_i > 0$ by combining outputs of each rule to a global target $V(s^k)$.
Step#5: Select the event with the highest reward/q value as the maximum firing strength.

Step#6: develop a reward or decrement for the fault identified by the FQL predictor as per the reward function.
Step#7: Evaluate temporal difference (TD) error.

$$\Delta Q = r + \gamma V(y^{k+1}) - Q(y^k, a(y^k)) \tag{9.2}$$

Where $\gamma \in (0,1]$ is the discount factor. For faster learning, eligibility trace mechanism [new ref] with a decay parameter $\xi \in (0,1]$ has been used.

$$\begin{aligned} e(i,a) \quad & \leftarrow e(i,a) \times \gamma \times \xi + \frac{\beta_i}{\sum \beta_i} && \text{for } a = a_{Iden} \\ & \leftarrow e(i,a) \times \gamma \times \xi && \text{for } a \neq a_{Iden} \end{aligned} \tag{9.3}$$

Step#8: Update rule specific $w(i, y)$ values as per Eq. (9.4) and the eligibility traces $e(i, y)$.

$$w(i, q_i^\dagger) \leftarrow w(i, q_i^\dagger) + \eta \, \Delta Q \frac{\alpha_i(cl)}{\psi} \tag{9.4}$$

Step#9: Decrease exploration rate as more at the starting and less at end.
Step#10: Estimate % success rate.

$$\%\text{success} = \frac{\text{success}}{\text{success} + \text{failure}} \times 100\% \tag{9.5}$$

Step#11: Again follow steps 1 to step 9 till q values converge, that is, $\Delta q \rightarrow 0$ or becomes a small enough value.

4.3 Data collection

The historical data for ELF has been obtained from GEFCom2012 [11], which has hourly data base for the year 2004–08 for the 20 different zones. Collected all dataset have been rearranged according to the proposed case studies in this chapter. Total three case studies have been proposed in this chapter: (1) hourly ELF, (2) weekly load forecasting, and (3) monthly load forecasting. To examine the property of large variation of the dataset, a statistical analysis is presented in Tables 9.3 and 9.4 for the dataset 2004–05 and 2006–07, respectively. In the statistical health check, minimum value (Min), maximum value (Max), mean, and standard deviation (STD) have been evaluated, which shows a fair consistency in the recoded data samples. Moreover, per hour/per month recorded data values have been represented in Figs. 9.7–9.18 for the month of January to December, respectively.

TABLE 9.3 Representation of statistical health check for 2004–05.

	Statistical information for year 2004				Statistical information for year 2005			
	Min	Mean	Max	STD	Min	Mean	Max	STD
January	9,859	23,183	39,584	5669.09	7,319	19,902	40,205	7327.709
February	12,147	20,890	34,609	4044.304	11,601	20,118	34,067	4355.808
March	9,168	15,824	30,858	3611.246	10,355	18,552	34,387	4499.196
April	9,129	14,756	24,261	2913.201	9,240	14,396	23,662	2648.073
May	9,110	16,292	27,725	4337.436	8,577	14,189	22,300	2767.306
June	9,500	17,402	30,192	4886.232	9,234	18,484	33,117	5720.411
July	10,727	20,750	33,436	6169.704	10,490	22,211	38,757	6699.028
August	9,379	18,614	32,792	5449.67	10,680	22,024	36,747	6710.005
September	8,774	15,448	25,944	3649.064	9,689	16,946	29,135	4798.275
October	8,688	13,961	19,176	2266.932	9,090	14,903	25,467	2879.581
November	9,258	16,008	25,878	3155.077	9,341	16,786	31,111	4231.141
December	11,231	21,292	44,869	5441.056	14,232	23,736	34,793	4236.08

TABLE 9.4 Representation of statistical health check for 2006–07.

	Statistical information for year 2006				Statistical information for year 2007			
	Min	Mean	Max	STD	Min	Mean	Max	STD
January	12,247	19,667	32,933	3810.579	10,906	22,215	40,641	5651.379
February	11,955	20,647	33,142	4184.226	11,976	25,791	45,547	6113.848
March	8,905	17,301	29,144	3819.861	9,744	17,429	33,749	4307.368
April	8,751	14,027	23,890	2559.555	9,229	16,487	28,706	3627.817
May	9,519	14,304	19,578	2294.358	9,602	16,869	30,486	4453.892
June	10,231	18,774	35,847	5493.094	10,683	20,192	35,313	6144.288
July	10,302	23,126	38,393	7135.041	10,905	21,920	38,656	6672.797
August	10,846	22,618	39,363	6749.775	12,309	25,649	43,486	7768.807
September	9,846	15,705	27,645	3652.058	9,846	18,636	35,512	5952.991
October	8,346	15,797	25,949	2908.005	9,828	16,133	29,407	3844.932
November	10,006	17,366	29,851	3840.226	10,431	18,424	32,055	3612.138
December	10,343	20,255	38,604	4525.994	10,821	21,272	34,938	4399.49

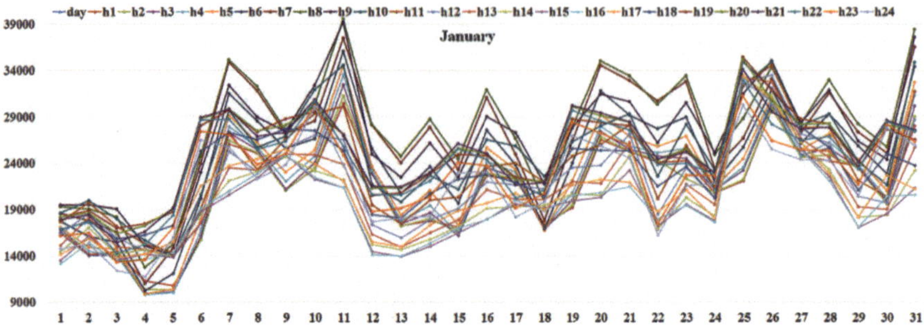

FIGURE 9.7 **Hourly data representation for January.**

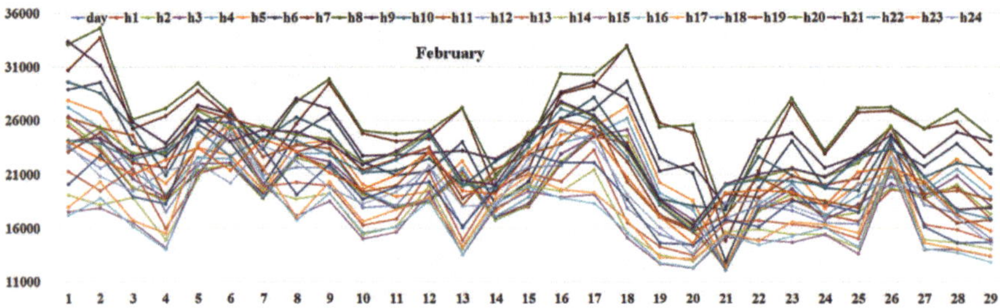

FIGURE 9.8 **Hourly data representation for February.**

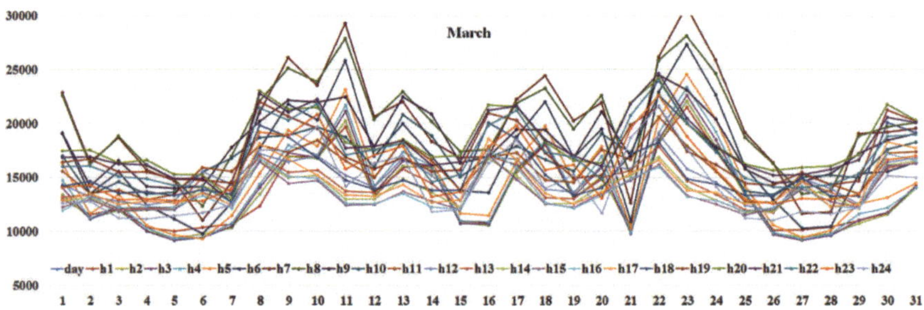

FIGURE 9.9 **Hourly data representation for March.**

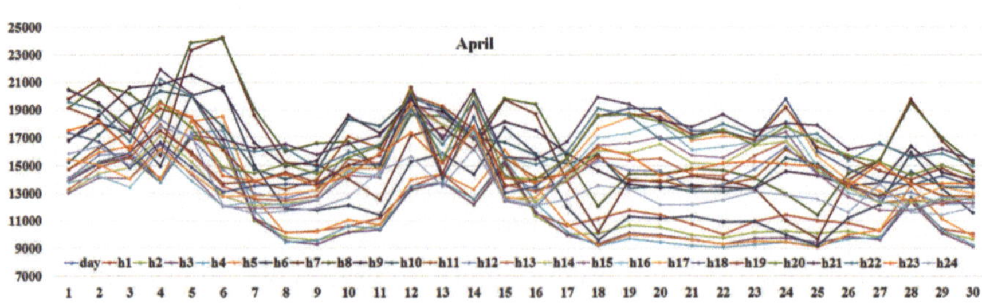

FIGURE 9.10 **Hourly data representation for April.**

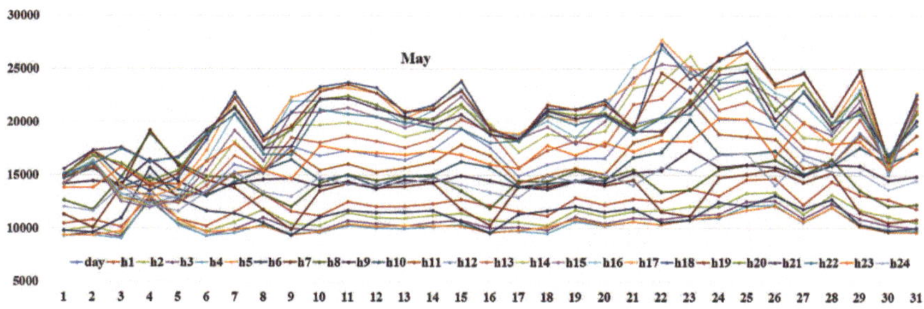

FIGURE 9.11 Hourly data representation for May.

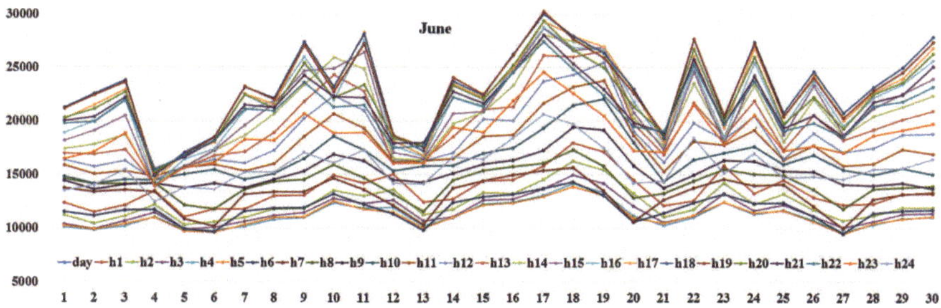

FIGURE 9.12 Hourly data representation for June.

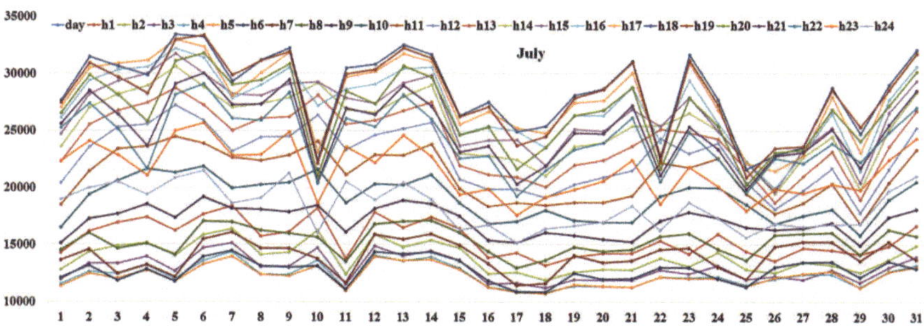

FIGURE 9.13 Hourly data representation for July.

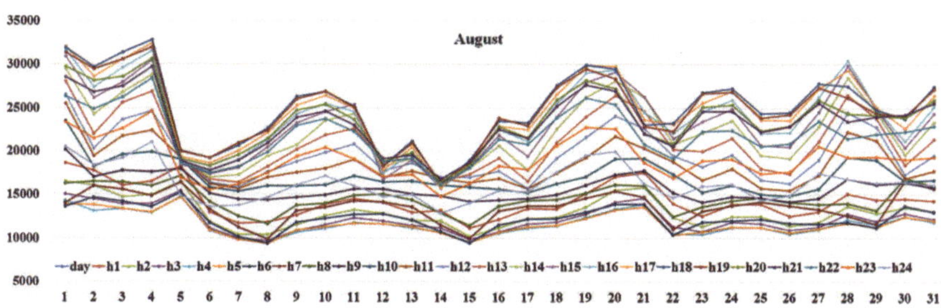

FIGURE 9.14 Hourly data representation for August.

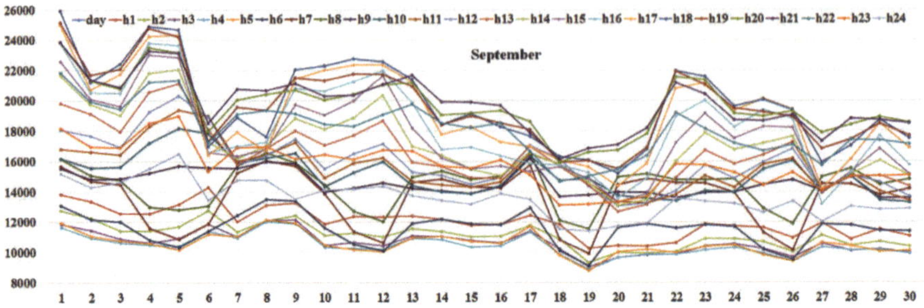

FIGURE 9.15 Hourly data representation for September.

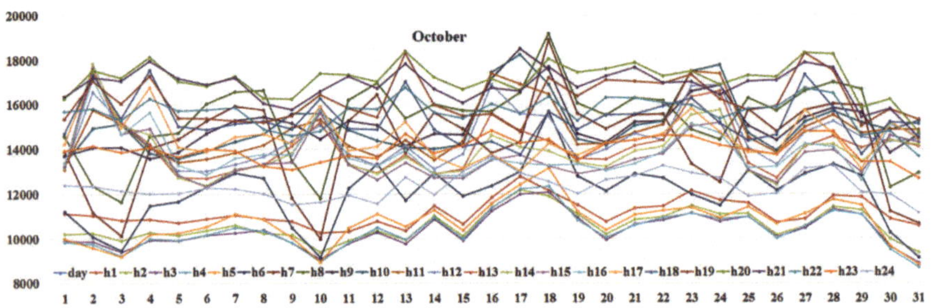

FIGURE 9.16 Hourly data representation for October.

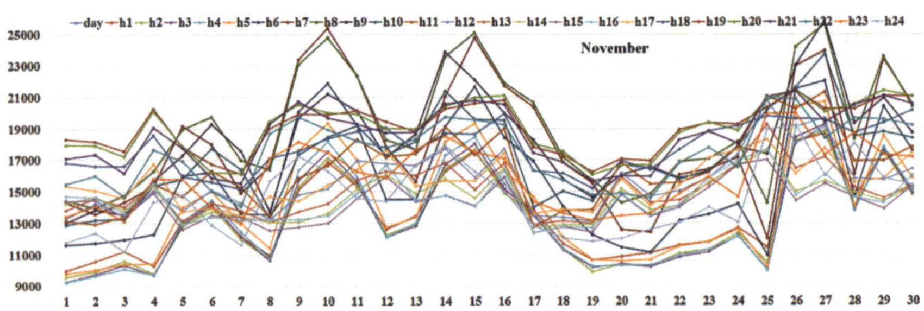

FIGURE 9.17 Hourly data representation for November.

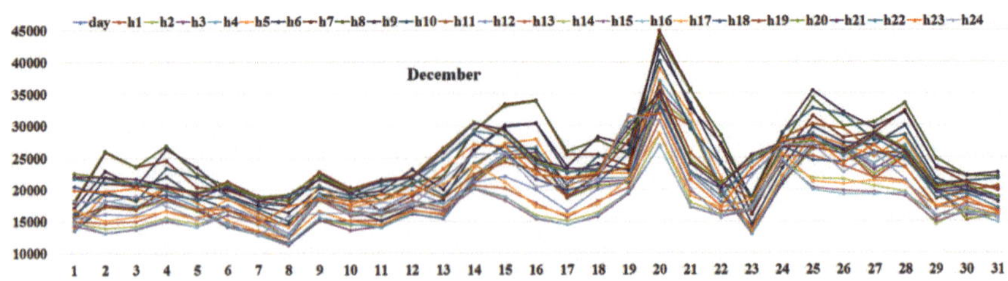

FIGURE 9.18 Hourly data representation for December.

5 Case studies: performance evaluation

The main aim of this section is to represent the obtained results from proposed approach and explained them in brief. The explanation of the obtained results is represented according to the analyses, which are performed into four different case studies. These case studies are: (1) monthly-ahead load forecasting, (2) week-ahead load forecasting, (3) day-ahead load forecasting, and (4) hour-ahead load forecasting. These four case studies are formulated with consideration of its distinct and valuable application in the power system as system planning, maintenance scheduling, unit commandment, economic load dispatch management, OPF analysis, automatic generation control, etc., as mentioned in Section 2. The dataset used in this study are taken for the year 2004–07, which are represented in Table 9.5. The obtained results from each case study have been summarized in Table 9.6. The value of MAPE presented in the Table 9.6 for each case is the average value of that particular case study. The detailed analysis of each case study has been represented separately in sub-sequence section as given further.

TABLE 9.5 Dataset used for FQL based FRL model formation for ELF.

Years	Monthly ELF	Weekly ELF	Daily ELF	Hourly ELF
Data sample (2004 Data)	12 samples	52 samples	366 samples	8784 samples
Data sample (2005 Data)	12 samples	52 samples	365 samples	8760 samples
Data sample (2006 Data)	12 samples	52 samples	365 samples	8760 samples
Data sample (2007 Data)	12 samples	52 samples	365 samples	8760 samples

TABLE 9.6 FQL based FRL performance for ELF: MAPE value.

Performance phase	Monthly ELF	Weekly ELF	Daily ELF	Hourly ELF
Training phase (2004–05 Data)	2.7374	5.502	4.2504	9.0158
Testing phase (2006 Data)	4.6525	**6.321**	5.5663	9.5177
Testing phase (2007 Data)	4.4785	**7.811**	5.4154	10.8252

5.1 Case study#1: month-ahead forecasting

Here, the final obtained results for month-ahead forecasting model are represented. Results are obtained by average value of the available hourly dataset then monthly average value is used for the model training as well as testing phase and hence results for training as well as testing phase are represented in Table 9.7. Here, multiple testing are performed to make sure the adoptability of the proposed model that testing for each case must be within a standard limit of MAPE. According to the [12–14], if MAPE is less than 10% then model is more acceptable condition. In this case study, highest value of MAPE is 4.25 for training phase and 5.89–5.53 for testing phase by using dataset of 2006 and 2007, respectively, which is more acceptable condition.

5.2 Case study#2: week-ahead forecasting

Here, in this section, week-ahead forecasting results in brief have been represented. The results are tabulated in the Table 9.8 for each week of each month for the year of 2006 and 2007. Presented results are during testing phase of the proposed model. The results have been analyzed by average value of the available hourly dataset information of week value then weekly average value is used for the model training as well as testing phase and hence results are represented in Table 9.8. Here, multiple testing are performed to make sure the adoptability of the proposed model. In this case study, the MAPE is less than 10% which shows the model's higher acceptable condition. In this case study, highest value of MAPE

TABLE 9.7 FQL based FRL performance for case study#1: MAPE for Month-ahead LF.

Month	2004–05	2006	2007
January	1.9814	3.8946	4.0558
February	2.5616	3.0533	5.2405
March	2.4510	4.0652	4.5625
April	4.2501	3.9895	3.6598
May	3.003	4.7525	4.0528
June	2.8052	4.7995	5.0851
July	2.6501	3.8635	4.9795
August	2.6921	5.8540	3.9233
September	2.2665	4.5676	4.6775
October	2.0845	5.7725	2.9895
November	2.9823	5.8937	5.5340
December	3.1205	5.3246	4.9812
Average	2.7374	4.6525	4.4785

TABLE 9.8 FQL based FRL performance for case study#2: MAPE for week-ahead LF.

	Year 2006 data based					Year 2007 data based				
	Week1	Week2	Week3	Week4	Average	Week1	Week2	Week3	Week4	Average
January	4.562	10.235	6.351	7.929	7.269	11.303	5.193	11.503	10.026	9.506
February	5.230	4.738	5.743	13.460	7.293	7.531	5.889	7.8012	6.9868	7.052
March	4.923	2.895	5.605	8.881	5.576	6.115	4.952	9.1173	6.8479	6.758
April	5.925	6.365	4.064	7.225	5.895	4.887	6.035	11.896	8.047	7.716
May	8.832	5.320	6.754	6.114	6.755	3.277	5.892	9.1266	6.087	6.096
June	2.705	4.687	5.980	7.242	5.154	5.854	6.889	9.4055	9.2706	7.855
July	3.662	7.892	4.534	5.649	5.434	4.025	6.078	7.3382	8.2321	6.418
August	4.751	8.052	6.193	5.842	6.21	6.829	7.351	8.2238	11.681	8.521
September	5.527	7.540	5.531	6.299	6.224	8.637	9.380	9.035	10.459	9.378
October	8.150	3.795	5.042	5.522	5.627	7.486	6.005	8.7364	6.6871	7.229
November	9.756	7.689	8.462	7.842	8.437	6.035	8.538	11.562	7.9472	8.521
December	3.558	5.892	7.491	6.985	5.982	4.881	9.890	10.282	9.6748	8.682
Average	5.632	6.258	5.979	7.416	**6.321**	6.405	6.841	9.502	8.496	**7.811**

varies from 8.462 to 13.460 for testing data of 2006 and 9.89 to 11.89 for testing data of 2007 and the average MAPE of all weeks for testing phase are 6.321 and 7.811 for testing dataset of 2006 and 2007, respectively. After analysis of the Table 9.8, it is concluded that proposed FRL approach is more acceptable for ELF.

5.3 Case study#3: day-ahead forecasting

For the day-ahead ELF, the average value of recorded data of a day is used for the further analysis. So in this way, total 366 and 365 data points are used for the year 2004 and 2005, respectively, which is used for the training phase of the FQL based FRL model and daily data points collected for year 2006 and 2007 have been used for the testing phases. The detailed results have been shown in Table 9.9. In the Table 9.9, it is clearly mentioned that maximum and minimum MAPE vary from 2.5456–11.529 for 2006 dataset and 1.7867–10.7182 for 2007 and the average values of MAPE are 5.56 and 5.414 for testing data1 and testing data2, respectively. Moreover, overall average value of evaluated MAPE is 5.4908, which shows the proposed approach outperformance.

5.4 Case study#4: hour-ahead forecasting

The main aim of this case study is to validate the proposed model acceptability and adaptability at a globalized manner for the ELF problem that proposed model has outperformed the performance for month-ahead, week-ahead, and day-ahead forecast of electric load. As in power system domain, the forecasting accuracy generally, decrease with respect to the decrease in forecasted time interval. Hence, forecasted accuracy for month-ahead is higher than week-ahead and so on. So, forecasting accuracy in the case of hour-ahead may be very less. So in this case study, it is validated that the forecasting performance of the proposed FQL base FRL approach has outperformance acceptability and it can be used for the online forecasting problem as well.

For the further validation of the proposed approach, the hour-ahead results are represented in Table 9.10 for each day from Monday to Sunday and then the

TABLE 9.9 FQL based FRL performance for case study#3: MAPE for day-ahead LF.

	Monday	Tuesday	Wednesday	Thursday	Friday	Saturday	Sunday	Average
2006 Data	11.5291	4.6576	2.7602	2.5456	3.0420	5.6408	8.7885	5.5663
2007 Data	10.7182	3.9123	3.0534	1.7867	2.8526	6.5042	9.0804	5.4154
Average	11.1237	4.285	2.9068	2.1662	2.9473	6.0725	8.9345	5.4908

TABLE 9.10 FQL based FRL performance for case study#4: MAPE for hour-ahead LF.

	Monday	Tuesday	Wednesday	Thursday	Friday	Saturday	Sunday	Average
hour1	9.3936	8.6582	8.5102	10.789	8.3259	9.4671	8.1332	9.0396
hour2	8.7333	8.3813	6.8881	6.9553	6.1244	8.7983	9.185	7.8665
hour3	9.2807	6.253	10.869	9.9225	9.5578	10.324	8.3819	9.227
hour4	7.2896	9.7602	10.921	9.4221	10.227	10.34	9.8344	9.6849
hour5	10.308	6.3306	7.7362	14.665	10.177	9.9815	7.8535	9.5788
hour6	8.2082	14.269	13.158	13.313	14.274	12.632	12.119	12.568
hour7	5.1172	8.758	14.382	10.117	8.7684	10.817	5.7707	9.1043
hour8	9.2355	12.532	10.558	6.4623	6.9455	14.455	12.735	10.418
hour9	9.2754	6.7652	6.1005	9.4466	8.4279	11.508	6.6977	8.3173
hour10	9.3157	7.4705	7.0562	11.565	6.7412	10.813	8.0391	8.7144
hour11	12.789	10.488	11.91	9.1726	12.083	11.63	5.5392	10.516
hour12	6.6381	8.1374	11.185	8.1613	11.592	10.517	9.8636	9.4421
hour13	12.896	7.9066	6.1011	13.846	10.82	11.248	6.1623	9.8543
hour14	11.448	8.0058	8.3963	10.18	7.4649	7.8974	10.244	9.0909
hour15	7.3692	14.928	13.544	14.102	11.685	12.816	14.123	12.652
hour16	9.1758	11.88	8.1929	9.719	11.282	9.1976	6.7483	9.4565
hour17	6.9785	12.17	7.134	12.099	12.314	9.4455	13.658	10.543
hour18	12.724	6.9429	9.2904	5.7626	6.7251	10.843	8.2714	8.6513
hour19	10.862	8.6539	8.5741	6.8729	7.3283	8.2719	9.7724	8.6194
hour20	10.159	8.0818	10.274	7.2959	9.6859	8.8528	8.0451	8.9135
hour21	13.251	12.427	5.0751	12.688	6.5916	10.84	14.859	10.819
hour22	10.469	10.018	9.3429	10.367	11.611	7.5811	7.1357	9.5035
hour23	5.0984	9.3004	9.4939	5.7181	9.8905	5.87	12.293	8.2378
hour24	7.6222	6.8896	6.9862	10.796	7.3575	6.4011	7.2009	7.6076
Average	9.3182	9.3753	9.2366	9.9766	9.4167	10.0228	9.2777	9.5177

average value for all day as well as all hours have been tabulated in this table. The MAPE value for each day (Monday to Sunday) for each hour case is varied from 9.185 to 14.928 and the overall average value for hour-ahead forecasting model is 9.5177, which is within highly acceptable condition. A forecasted model is highly acceptable if MAPE <10. So, in this study, the MAPE is less than the specific value, which validates that the model acceptability position is very high.

6 Conclusion and future work

In this chapter, an intelligent model for ELF has been proposed by using FQL based FRL model. Proposed model's performance and its adaptability have been validated by using four different case studies of month-ahead, week-ahead, day-ahead, and hour-ahead forecasting of the load. In the all forecasting case studies, the electric load has predicted the demand of month-ahead, by a step of 1 month. Similarly, in week-ahead forecasting, the electric load has predicted the demand of week-ahead, by a step of 1 week. In day-ahead forecasting, the electric load has predicted the demand of week-ahead, by a step of 1 day. And finally, in four-ahead forecasting, the electric load has predicted the demand of hour-ahead, by a step of 1 h. The proposed FQL based FRL model for all case studies have been developed by using historical load profile dataset of hourly recorded. To generate the monthly, weekly, and daily data profile, the recorded hourly data have been used. To generate the monthly data, recorded hourly readings of a month are used and an average value has been calculated. Similarly, for generating weekly data, hourly reading of a week is used to get the average value and so on. After generating all data sets for all case studies, FRL model has been designed and trained by using historical dataset of years 2004–05. The trained model has been tested multiple times by using two different dataset of year 2006 and 2007. The obtained results during training and testing phase of the proposed approach show outperform. The proposed approach may be implemented for online ELF for the different power system applications.

References

[1] T. Ahmad, H. Zhang, B. Yan, A review on renewable energy and electricity requirement forecasting models for smart grid and buildings, Sustain. Cities Soc. 55 (2020) 102052, https://doi.org/10.1016/j.scs.2020.102052.

[2] M. Bourdeau, et al., Modeling and forecasting building energy consumption: A review of data-driven techniques, Sustain. Cities Soc. 48 (2019) 101533, https://doi.org/10.1016/j.scs.2019.101533.

[3] T. Ahmad, H. Chen, A review on machine learning forecasting growth trends and their real-time applications in different energy systems, Sustain. Cities Soc. 54 (2020) 102010, https://doi.org/10.1016/j.scs.2019.102010.

[4] R. Khalid, N. Javaid, A survey on hyperparameters optimization algorithms of forecasting models in smart grid, Sustain. Cities Soc. 61 (2020) 102275, https://doi.org/10.1016/j.scs.2020.102275.

[5] S. Fathi, R. Srinivasan, A. Fenner, S. Fathi, Machine learning applications in urban building energy performance forecasting: A systematic review, Renew. Sustain. Energy Rev. 133 (2020) 110287, https://doi.org/10.1016/j.rser.2020.110287.

[6] H. Malik, R. Sharma, Transmission line fault classification using modified fuzzy Q learning, IET Generat. Transm. Distrib. 11 (16) (2017) 4041–4050, 10.1049/iet-gtd.2017.0331.

[7] H. Malik, R. Sharma, S. Mishra, fuzzy reinforcement learning based intelligent classifier for power transformer faults, ISA Trans. 101 (2020) 390–398, https://doi.org/10.1016/j.isatra.2020.01.016.

[8] A. Kukker, R. Sharma, H. Malik, Reinforcement learning based genetic fuzzy classifier for transformer faults, IETE J. Res. (in press), doi: https://doi.org/10.1080/03772063.2020.1732844

[9] N.K. Nandan, R. Sharma, H. Malik, Solving nonconvex economic thermal power dispatch problem with multiple fuel system and valve point loading effect using fuzzy reinforcement learning, J. Intell. Fuzzy Syst. 35 (5) (2018) 4921–4931, doi: 10.3233/JIFS-169776.

[10] H. Malik and S. Mishra, Application of fuzzy Q learning (FQL) technique to wind turbine imbalance fault identification using generator current signals, in: Proceedings of IEEE Seventh Power India International Conference, Bikaner India, 2016, pp. 1–6, doi: 10.1109/POWERI.2016.8077283

[11] GEFCom2012 Load Forecasting Data. (2020). Available from: http://blog.drhongtao.com/2016/07/gefcom2012-load-forecasting-data.html. Accessed 30.08.2020

[12] A Azeem, et al., k-NN and ANN based deterministic and probabilistic wind speed forecasting intelligent approach, J. Intell. Fuzzy Syst. 35 (5) (2018) 5021–5031, doi: 10.3233/JIFS-169786.

[13] P Arora, et al., Wind energy forecasting model for northern-western region of India uing decision tree and MLP neural network approach, Interdiscip. Environ. Rev 19 (1) (2018) 13–30, doi: 10.1504/IER.2018.089766.

[14] A.K. Yadav, et al., A novel hybrid approach based on relief algorithm and fuzzy reinforcement learning approach for predicting wind speed, Sustain. Energy Technol. Assess. 43 (2021), doi: 10.1016/j.seta.2020.100920 In press.

Chapter 10

Intelligent Data Analytics for Battery Health Forecasting Using Semi-Supervised and Unsupervised Extreme Learning Machines

1 Introduction

Batteries are utilized for energy source and storage in several fields such as in electric vehicle (EV), communication, navigation, emergency back-up, aviation, and many more. According to the global patent report, the growth of research for the battery storage system can be analyzed based on the published number of patents by patent office globally during 2016–18 as shown in Fig. 10.1. Fig. 10.1 shows that the big player in the battery domains are from only three regions such as the United States patent and trademark office (US-PTO), European patent office (EPO), Chinese patent office (CPO), and rest of the world. The 9856 total number of patents have been granted by the patents offices in the world during 2016–18 for the R&D of battery energy storage, in which CPO contribute the highest number of patents around 43% followed by US-PTO (28% plus) and EPO (6.3% plus). Around 5368 patents have been filed within the Asian region, followed by North-America of around 2803 patents. According to the technology growth of battery storages, different types of inventions are going on day-by-day and a huge number of the invention are being filed to the patent offices throughout the globe. The seven different types of battery storage inventions are going on in the research domain as shown in Fig. 10.2, such as DIB (dual-ion battery), Li-S (lithium-sulfur battery: LSB), Mg-Na (magnesium-based battery: MBB), Na-Ion (sodium-ion battery: SIB), K-Air (potassium-air battery: PAB), Ni-Zn (nickel-zinc battery: NZB), and Al-Air (aluminum-air battery (AAB). Therefore, the use of a particular type of battery storage depends on the different types of its properties (i.e., storage capacity, cost, performance, safety, charging/discharging, applicability, and level of technology, etc.) as shown in Fig. 10.3.

According to Fig. 10.3, among the different available technologies of battery, lithium-ion battery (LIB) is utilized widely in the world due to its superior

Intelligent Data-Analytics for Condition Monitoring. http://dx.doi.org/10.1016/B978-0-323-85510-5.00010-7

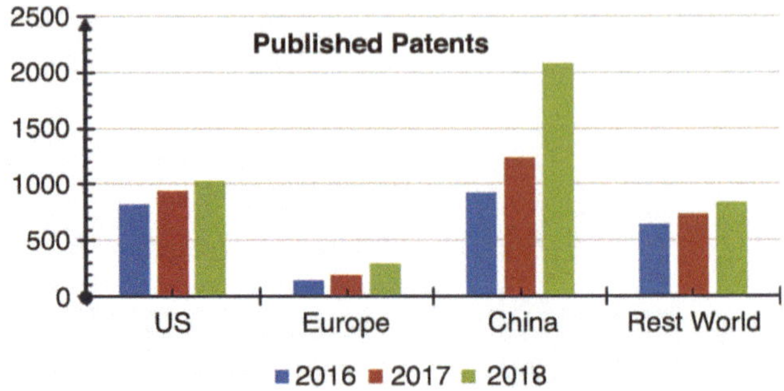

FIGURE 10.1 Region-wise published patents in the area of battery storage globally [1].

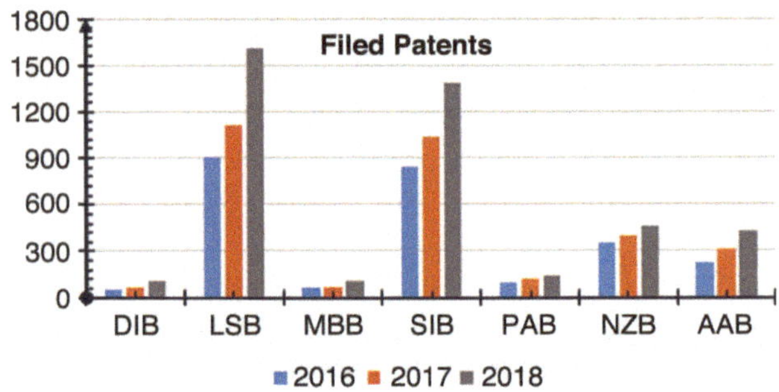

FIGURE 10.2 File patents in the ongoing new research of the battery storage [1].

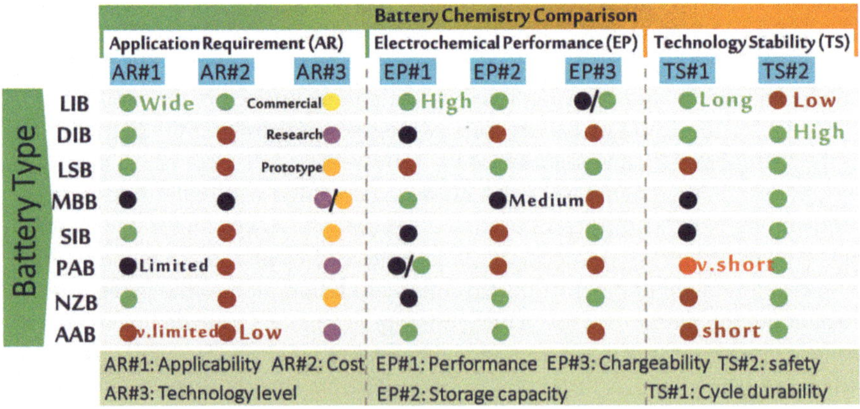

FIGURE 10.3 Different types of properties of the battery chemistries [1].

properties, namely long life time, high energy density, light weight, low self-discharge rate, the clean environmental effect, and high output voltage [2]. Moreover, the forecasted global LIBs market is around 629.22 USD$ billion by 2025, which includes all cumulative sales globally. The global market demand

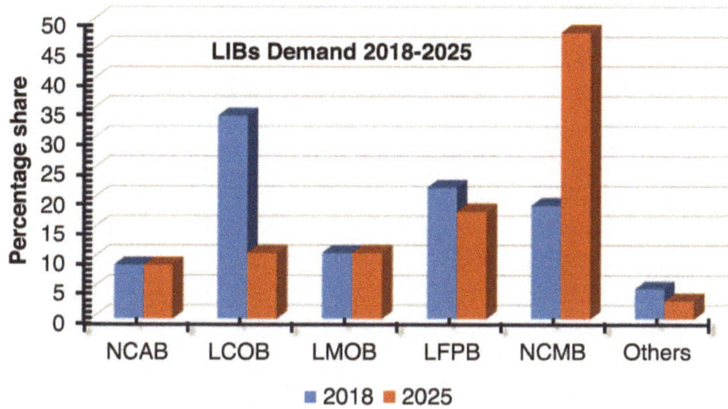

FIGURE 10.4 **LIBs market demand globally [1].**

for LIBs (i.e., lithium cobalt oxide battery-LCOB, lithium nickel cobalt aluminum oxide battery-NCAB, lithium manganese oxide battery-LMOB, lithium iron phosphate battery-LFPB, and lithium nickel manganese cobalt oxide battery-NCMB) is shown in Fig. 10.4, which represents the comparative growth rate of all types LIBs from 2018 to 2025. Fig. 10.4 shows that the sharing part of NCMB will be enhanced around 48%, which will be the remarkable changes in LIB industry. Moreover, according to the industry, the demand for LIBs varies with respect to the applications.

The repetitive charge/discharge condition deteriorates the battery performance gradually and it may lead to drastic failure. For example, NASA's robotic spacecraft (Mars Global Surveyor) became a failure due to its battery failure in 2006 [2]. Boeing 787 Dreamliner was indefinitely grounded due to its battery failure in 2013 [3]. The fire incidents due to over-charging of Li-ion battery packs in Singapore have been doubled in 2017 as compared to 2016 [4]. As per Singapore Civil Defense Force (SCDF) report, the number of fires involving e-scooters rose by 350% in 2017 as compared to 2016. As of February 1, 2019, 241 air/airport building incidents happened as a result of the failure of Li-ion batteries [5]. For a system like solar Photovoltaic which utilizes the battery for storage, battery cost could vary from 23% to 55% cost of the total system cost [6]. Therefore, given the safety and high cost of the storage system, accurate- and reliable health assessment of battery is a necessity. So, the present day research focus is on battery health assessment in terms of battery capacity fade estimation as well as estimation of remaining-useful-life (RUL) to address the necessity of charging/discharging control and safety management.

In this regard, health assessment and RUL estimation can be performed easily by using capacity or internal resistance as an indicator. However, the online measurement of these two variables is very difficult. This is because the battery's internal state variables are inaccessible to sensors under operational conditions and direct and accurate measurement of internal parameters is very costly, which makes the online condition monitoring utilizing these parameters ineffective. Therefore, non-intrusive extraction of health indicators from the commonly measured variables, which may reflect the health condition of the battery in online mode has become significant.

In this chapter, a health indicator (HI) is extracted by using charge and discharge condition measurements. After extraction, optimization is performed to enhance the linear correlation between extracted HI and actual health condition of the battery. Now, transformed HI is used to find out the health condition instead of the actual condition, which allows assessing the online health estimation of the battery non-intrusively.

The organization of this chapter is comprised of four sections. Section 1 represents the introduction and motivation part of the study. Section 2 explains the methodology part, which includes the experimental dataset collection, proposed approach framework, HI extraction, and its performance evaluation methods, RUL prediction techniques (SELM, SSELM, and USELM) and involved steps in the proposed approach. Section 3 presents the results and its discussion and experimental validation of all steps as mentioned in the proposed approach. Finally, Section 4 presents the conclusion of the study.

2 Methodology

2.1 Data collection for study

For the experimentation validation of the proposed approach, the LIBs datasets collected from NASA [7] are considered to verify the presented scheme for HI extraction and multi-ELMs based RUL estimation. Dataset were collected from battery prognostic testbed 18650, which include commercially available rechargeable LIBs. Testbed includes voltmeters, sensors, programmable load, environmental chamber, DAS, electrochemical impedance spectroscopy (EIS), and PCI extension. The experiments on a set of four LIBs (namely no. #5, #6, #7, and #18) were performed through three different operational conditions (charging, discharging, and impedance measurement) at room temperature.

1. Charging process: Charging experiment was performed in a constant current (CC) mode at 1.5 A until the battery voltage reached 4.2 V and then continued in a constant voltage (CV) mode until the charge current dropped to 20 mA.
2. Discharging process: Discharge experiment was performed at a CC level of 2 A until the battery voltage fell to 2.7, 2.5, 2.2, and 2.5 V for batteries #5–#7, and #18, respectively.
3. Impedance measurement: Impedance experiment was performed through an EIS frequency sweep from 0.1 Hz to 5 kHz. Repeated charge/discharge experiment result in accelerated aging of the batteries while impedance experiment provide insight into the internal battery parameters that change as aging progresses.

Repetitive charging/discharging operation was stopped when the batteries reached end-of-life (EOL) criteria, which was 30% fade in rated capacity (from 2 to 1.4 Ah). The recorded dataset is utilized for HI extraction and multi-ELM based RUL estimation. The recorded qualities during both charging and discharging conditions have been represented in Figs. 10.5–10.9 and Figs. 10.10–10.14, respectively.

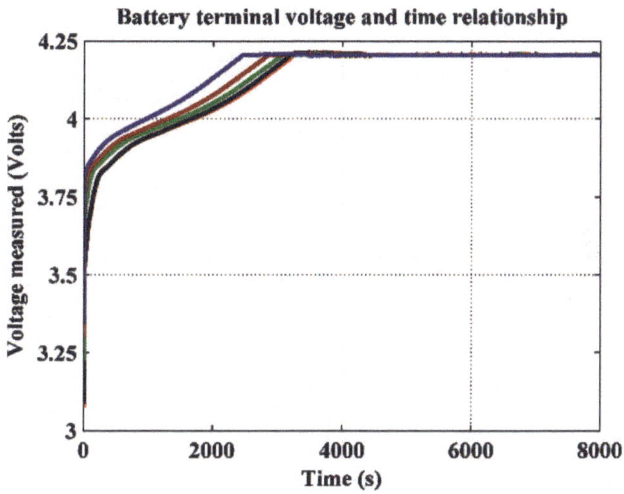

FIGURE 10.5 Terminal voltage versus time representation during charging condition.

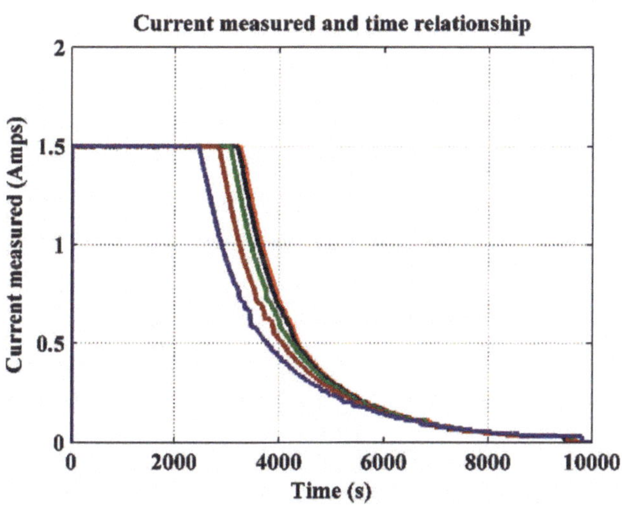

FIGURE 10.6 Current measured versus time representation during charging condition.

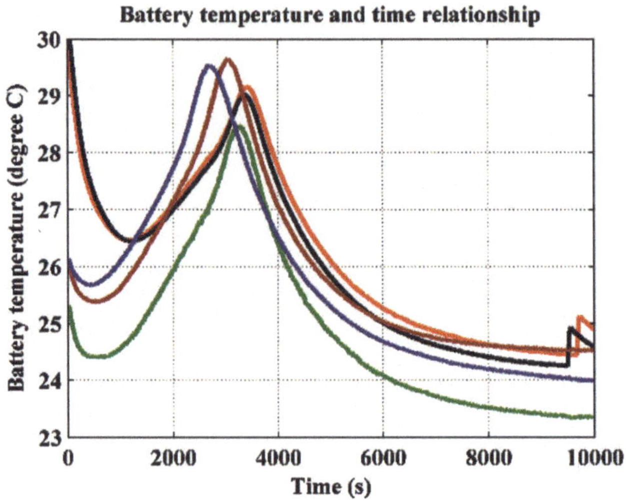

FIGURE 10.7 Battery temperature and time relationship.

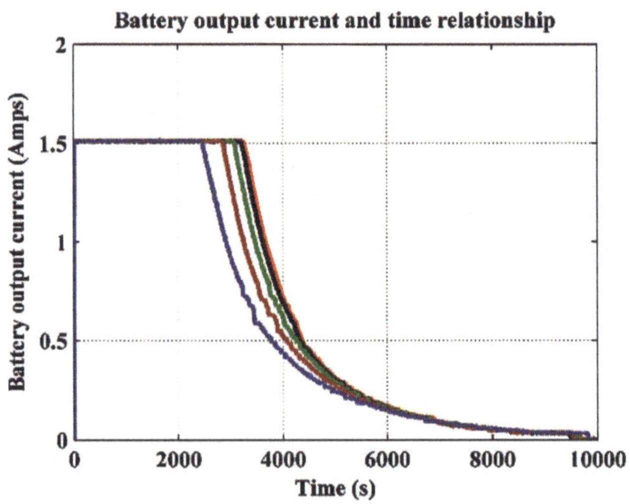

FIGURE 10.8 Output current versus time representation during charging condition.

2.2 Proposed approach framework

Fig. 10.15 represents the proposed approach and methodology, which includes a complete procedure for HI extraction, HI optimization, HI validation, and finally RUL evaluation for LIBs. This scheme utilized the raw experimental datasets for the extraction and optimization of HI. Initially, trends of each recorded parameters have been analyzed to extract the HI then a correlation is performed to map the extracted HI with battery capacity. The presented

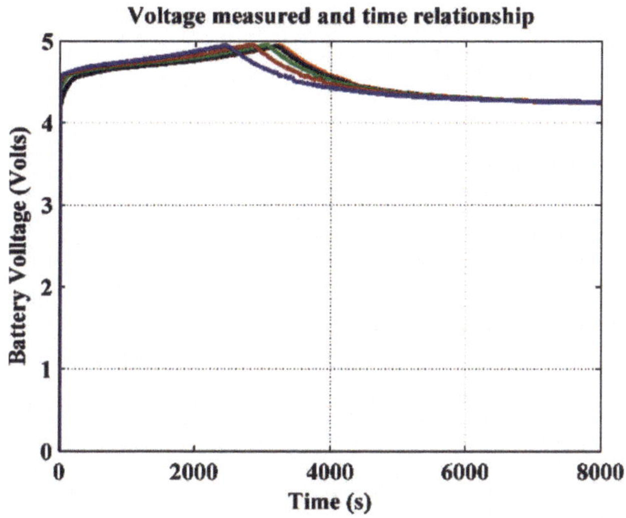

FIGURE 10.9 Voltage measured versus time representation during charging condition.

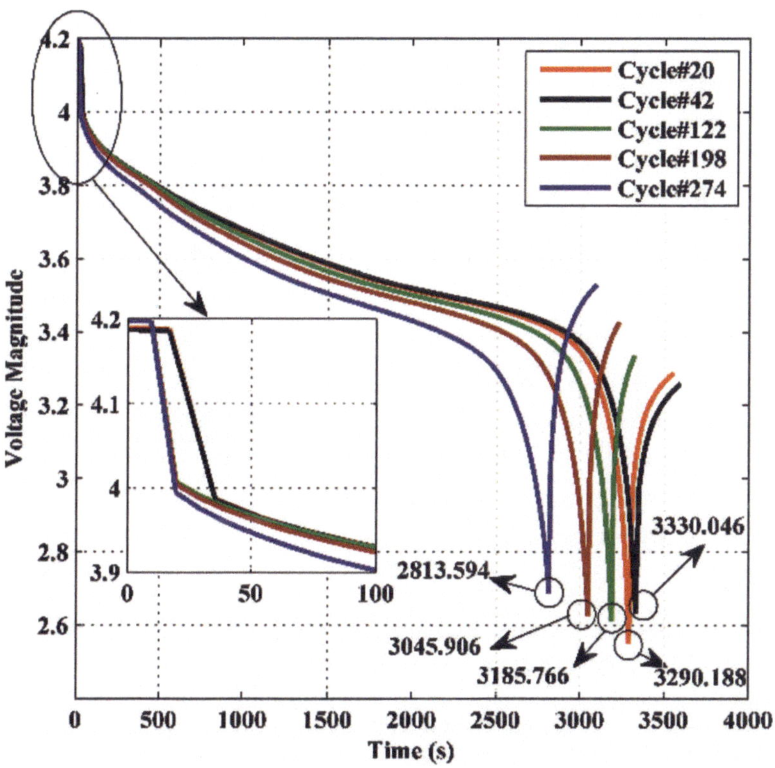

FIGURE 10.10 Voltage measured versus time representation during discharging condition.

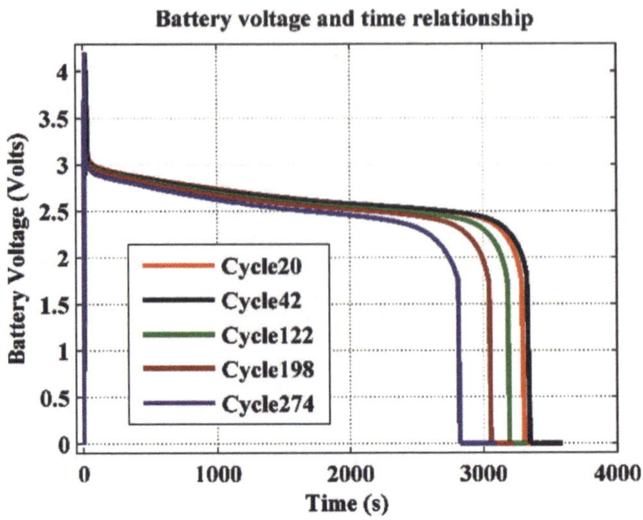

FIGURE 10.11 Voltage versus time representation during discharging condition.

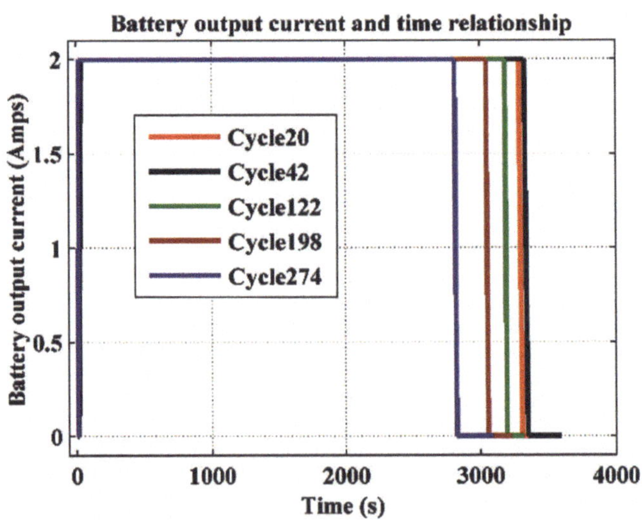

FIGURE 10.12 Output current versus time representation during discharging condition.

scheme is applicable for online non-intrusive health assessment of LIBs, which is not possible by measurement of R and capacity of battery directly. In subsection#1 (HI extraction), real-time dataset collection, and HI extraction are presented. For HI extraction, indirect HIs are extracted for the estimation of RUL for LIBs. After this, HI optimization is performed to enhance the correlation between extracted HI (eHI) and the actual capacity of the battery. Box-Cox

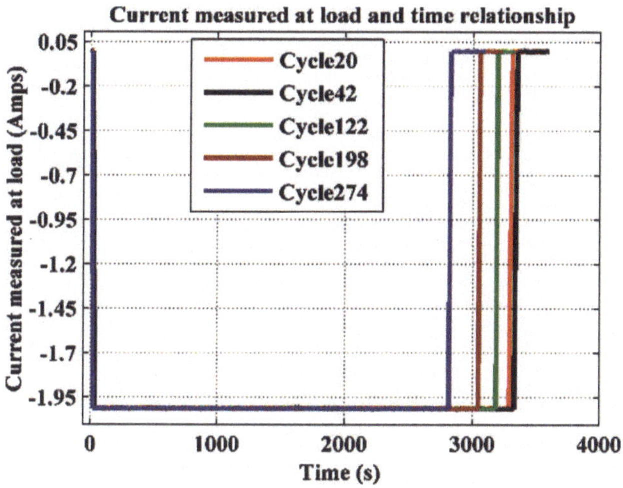

FIGURE 10.13 Current measured at load versus time representation during discharging condition.

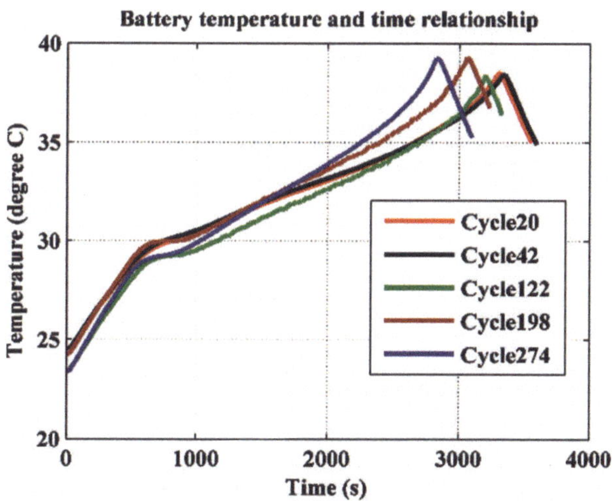

FIGURE 10.14 Temperature versus time representation during discharging condition.

transformation method (BCTM) used to enhance the correlation and its performance is validated by using Pearson & Sparceman correlation method. After getting optimized HI, the dataset is divided into two part for training and testing purposes. In Section 3, multiple ELMs are developed, trained, and tested one by one. The design and development for all ELMs approaches are based on Refs. [8,9]. The complete procedure is explained in detail in sub-sequence sections.

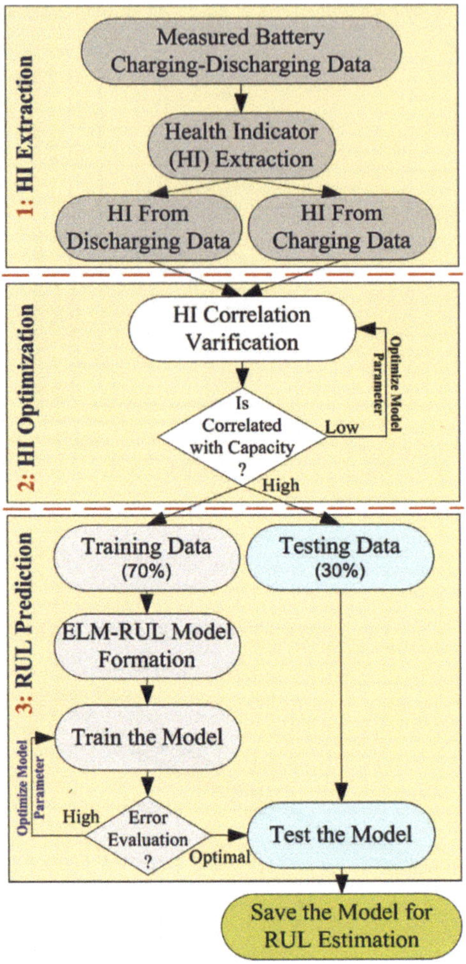

FIGURE 10.15 **Approach for HI and RUL estimation.**

2.2.1 Formulation of HI extraction and optimization

2.2.1.1 HI extraction

For online non-intrusive health assessment of LIB, HI attribute should be easy to obtain during the operating scenarios. For LIB, the operating time, temperature, voltage, and current are easy to measure under charging and discharging conditions during live operation of the battery. So, HI can be extracted from the charging as well as discharging dataset of the battery. After analyzing Fig. 10.10, it is concluded that this curve is very similar to the deterioration curve. Discharging condition voltage has also physical significance with the degradation of battery capacity [10]. The collected data samples during charging/discharging

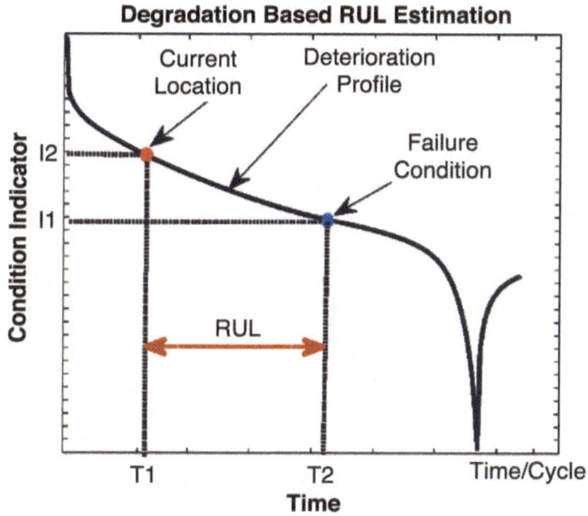

FIGURE 10.16 **Degradation based RUL estimation.**

conditions are not of the same size. So, this data may not be utilized directly for feature extraction and it needs to be processed. To resolve this problem, the same point at the equal interval for each cycle has been choosing as an indicator (Fig. 10.16) of battery health assessment as shown in Eq. (10.1).

$$TI_{(i)}\big|_{DVI} = \left|T_{(i)DV\,\max} - T_{(i)DV\,\min}\right| \tag{10.1}$$

where, $T_{(i)DV\,\max}$ = time at high voltage magnitude, $T_{(i)DV\,\min}$ = time at low voltage magnitude, and i = cycle number.

Now, a vector of these calculated TI for 168 cycle of discharging condition has been formulated as:

$$TI_{Raw} = [TI_{(1)}, TI_{(2)}, TI_{(3)}, \ldots\ldots, TI_{(i)}]^T \tag{10.2}$$

This vector [Eq. (10.2)] is used as a HI for RUL assessment of the battery.

2.2.1.2 HI optimization using Box-Cox transformation and its parameter identification

The curve obtained from vector of Eq. (10.2) has similar geometric characteristics of battery capacity, but it needs to be enhanced for a better correlation between them. For this, HI optimization is performed by Box-Cox (BC) transformation [11] and validated by Pearson correlation analysis (PCA) and Spearman correlation analysis (SCA) methods [12].

Generally, BC transformation is a family of power transformation. It modifies normally distributed dataset to a set of data that has approximately normal distribution. The BC is represented mathematically as:

$$TI_{Raw}(\lambda) = \begin{cases} \dfrac{TI_{Raw}{}^{\lambda} - 1}{\lambda} & if\,\lambda \neq 0 \\ \log(TI_{Raw}) & if\,\lambda = 0 \end{cases} \tag{10.3}$$

where, logarithm is the natural logarithm (log base e). The algorithm calls for finding the λ value that maximizes the log-likelihood function (LLF). Here, BC has been implemented by using R software then improvement of correlation between HI and battery capacity has been evaluated by PCA and SCA methods.

The PCA evaluates the linear dependence (r) (linear relationship) between TI and capacity which varies from +1 to −1 and represents the significant of correlation as $r = 1$, $r = 0$, and $r = -1$ for positive linear, no linear, and negative linear correlation, respectively. Pearson correlation coefficient (r):

$$r = \frac{\sum_{i=1}^{n}(TI_i - \overline{TI}) * (Ah_i - \overline{Ah})}{\sqrt{\sum_{i=1}^{n}(TI_i - \overline{TI})^2}\sqrt{\sum_{i=1}^{n}(Ah_i - \overline{Ah})^2}} \tag{10.4}$$

where, Ah = actual capacity, $\overline{Ah}$ = mean value of Ah.

SCA is a nonparametric measure of rank correlation using monotonic function (whether linear or not). A perfect SCA is represented in term of +1 or −1, if one variable is perfectly monotonic function to the other variable.

2.2.1.3 HI performance evaluation

Here, HI performance evaluation is performed by using root mean squared $(0 \leq \%RMSE \leq 0.5)$ and degree of fitness $(0 < R^2 \leq 1)$, which are evaluated as Eq. (10.5) and Eq. (10.6), respectively.

$$RMSE = \sqrt{\frac{\sum_{i=1}^{n}(Ah - \widehat{Ah})^2}{n}} \tag{10.5}$$

$$R^2 = 1 - \frac{\sum_{i=1}^{n}(Ah - \widehat{Ah})^2}{\sum_{i=1}^{n}(Ah - \overline{Ah})^2} \tag{10.6}$$

Where, $\widehat{Ah}$ = estimated capacity after transformation

2.2.2 RUL evaluation using multi-ELM

2.2.2.1 RUL prediction

The main component of a predictive maintenance strategy is to find out the RUL of equipment. RUL is the time length of healthy operation without/before repair/replacement of the equipment. So, equipment maintenance can be scheduled properly by estimation of RUL. Generally, RUL of LIB can be predicted by a model-driven method and data-driven method. The common way/models based on data-driven methods to estimate the RUL are: (1) similarity model, (2) survival model, and (3) degradation model, which depends on a different kind of dataset such as: (1) lifetime data, (2) run-to-failure histories dataset, and (3) known threshold value of specific condition data, respectively [13]. In this study, data type 3 is used to analyze the RUL of battery number 5, which can be easily understood graphically by Fig. 10.16.

2.2.2.2 Mathematical modeling of ELMs for RUL evaluation

The ELM approach can be implemented for the both type of problems supervised (labelled dataset) and unsupervised (unlabelled dataset). For the labelled dataset, standard ELM (SELM) approach is implemented whereas for the unlabelled dataset, unsupervised ELM (USELM) approach is implemented in a simple way which is similar to spectral clustering (SC) and Laplacian eigenmaps (LE). However, USELM can be implemented straightforward to handle any data without evaluating eigenvectors. Moreover, if data set is the mixture of labelled and unlabelled dataset then semi-supervised ELM (SSELM) is implemented. The SSELM is based on SS learning algorithms (i.e., Laplacian regularized least squares-LapRLS, Laplacian SVMs-LapSVMs, SS deep embedding-SSDE, SS neural networks-SSNNs). The design and implementation of SELM, SSELM, and USEM are based on Ref. [8] and its code developments are as per Ref. [9].

2.2.2.2.1 Mathematical modeling of SELM Let's assume training labelled data:

$$[S,T]=[s_i,t_i]_{i=1}^{N} \tag{10.7}$$

Where $s_i \in \mathbb{R}^{n_i}$, $t_i \in \mathbb{R}^{n_1}$ for classification and $t_i \in \mathbb{R}^{n_0}$ for regression problem, t_i = binary dimension vector.

The training procedure for SELM includes two general steps as follows:

Step#1) To construct the hidden layer using randomly generated neurons using:

Sigmoid function,

$$z(s,\lambda)=\frac{1}{1-\exp(-(a^T s+b))} \tag{10.8}$$

Gaussian function,

$$z(s,\lambda)=\exp(-b\|s-a\|) \tag{10.9}$$

Where, $\lambda=[a,b]$ = parameters and $\|\cdot\|$ = Euclidean norm.

In SELM, as per continuous probability distribution, parameters of hidden mapping function are generated randomly. In this step, hidden neurons map the training data into n_h feature space and represented as:

Output of hidden layer with respect to s_i,

$$h(s_i) \in \mathbb{R}^{1\times n_h} \tag{10.10}$$

Output weight (used to connect hidden layer & output layer),

$$\omega \in \mathbb{R}^{n_h \times n_O} \tag{10.11}$$

Then output of the network,

$$O(s_i) = h(s_i)\omega \tag{10.12}$$

Step#2) Optimize the output weights by minimizing the error

$$\min_{\omega \in \mathbb{R}^{n_h \times n_O}} \underbrace{\frac{1}{2}\|\omega\|^2}_{\text{---PART-1---}} + \underbrace{\frac{C}{2}\sum_{i=1}^{N}\|e_i\|^2}_{\text{-----PART-2-----}} \tag{10.13}$$

$$s.t. \quad h(s_i)\omega = t_i^T - e_i^T \tag{10.14}$$

Where, PART-1 = regularization term against over-fitting, $e_i \in \mathbb{R}^{n_O}$ = error of ith training sample, C = penalty factor for e.

Subtract the constraints form objective function, Eq. (10.13):

$$\min_{\omega \in \mathbb{R}^{n_h \times n_O}} J_{ELM} = \underbrace{\frac{1}{2}\|\omega\|^2}_{\text{--PART-1--}} + \underbrace{\frac{C}{2}\|T - P\omega\|^2}_{\text{---PART-2---}} \tag{10.15}$$

Where,

$$P = [h(s_i)^T, \ldots., h(s_N)^T]^T \in \mathbb{R}^{N\times n_h} \tag{10.16}$$

Now equating gradient of J_{ELM} with respect to ω to zero as:

$$\nabla J_{ELM} = \omega + CP^T(T - P\omega) = 0 \tag{10.17}$$

If $s > h_n$ (i.e., P has more row than column), the Eq. (10.15) will be:

$$\omega^* = \left(P^T P + \frac{L_{n_h}}{C}\right)^{-1} P^T T \tag{10.18}$$

where L_{n_h} = identity matrix.

if $s < h_n$ (i.e., P has more column than row), the Eq. (10.15) will be:

$$\omega^* = P^T \beta^* = P^T \left(PP^T + \frac{L_N}{C} \right)^{-1} T \tag{10.19}$$

Where,

$$\beta + C\left(T - PP^T \beta\right) = 0 \tag{10.20}$$

Therefore, the optimized output weight (ω) will be evaluated as per the number of training data samples (s_i) and hidden layer neurons (h_n) by using Eqs. (10.18) and (10.19).

2.2.2.2.2 Mathematical modeling of SSELM

The SSELM is helpful to analyze the mix dataset of both labelled and unlabelled dataset. Let's assume:

$$\text{Labelled dataset: } [S_l, T_l] = [x_i, t_i]_{i=1}^{l} \tag{10.21}$$

$$\text{Unlabelled dataset: } S_u = [x_i]_{i=1}^{l} \tag{10.22}$$

Where, l = labelled data samples and u = unlabelled data samples.

After modification of SELM of Eqs. (10.19) and (10.20), SSELM can be represented as:

$$\begin{aligned} &\min_{\omega \in \mathbb{R}^{n_h \times n_O}} \frac{1}{2}\|\omega\|^2 + \frac{1}{2} C_i \sum_{i=1}^{l} \|e_i\|^2 + \frac{\lambda}{2} Tr\left(H^T J H\right) \\ &s.t.\ \ h(s_i)\omega = t_i^T - e_i^T, \qquad i = 1,..,l \\ &\qquad O_i = h(s_i)\omega, \qquad i = 1,..,l+u \end{aligned} \tag{10.23}$$

Where, $J \in \mathbb{R}^{(l+u)\times(l+u)}$ = graph Laplacian obtained from both type of datasets, $H \in \mathbb{R}^{(l+u)\times n_O}$ = output matrix for i^{th} row which is $o(s_i)$, and λ = tradeoff parameter (TP), $C_i = \dfrac{C_o}{N_{t_i}}$ = penalty factor, C_o = parameter (user define), N_{t_i} = training sample.

After substituting the constraints, Eq. (10.23) is represented as:

$$\min_{\omega \in \mathbb{R}^{n_h \times n_O}} \frac{1}{2}\|\omega\|^2 + \frac{1}{2}\left\| C^{\frac{1}{2}}\left(\tilde{T} - P\omega\right)\right\|^2 + \frac{\lambda}{2} Tr\left(\omega^T P^T J P \omega\right) \tag{10.24}$$

Where, $\tilde{T} \in \mathbb{R}^{(l+u)\times n_O}$ = augmented training target for 1 rows ~ T_l (rest are zero).

Now, value gradient of Eq. (10.24) with respect to ω is

$$\nabla J_{SSELM} = \omega + P^T C\left(\tilde{T} - P\omega\right) + \lambda.P^T JP\omega \tag{10.25}$$

The output weight vector if $s > h_n$ is

$$\omega^* = \left(L_{n_h} + P^T CP + \lambda P^T JP\right)^{-1} P^T C\tilde{T} \tag{10.26}$$

The output weight vector if $s < h_n$ is

$$\omega^* = P^T \left(L_{l+u} + CPP^T + \lambda JPP^T\right)^{-1} C\tilde{T} \tag{10.27}$$

2.2.2.2.3 *Mathematical modeling of USELM*

Assumed again unlabeled dataset similar to Eq. (10.22): $S = [x_i]_{i=1}^{N}$.

Then formula for USELM is:

$$\min_{\omega \in \mathbb{R}^{n_h \times n_O}} \|\omega\|^2 + \lambda Tr\left(\omega^T P^T JP\omega\right) \tag{10.28}$$

Eq. (10.22) can be represented as:

$$\min_{\omega \in \mathbb{R}^{n_h \times n_O}} \|\omega\|^2 + \lambda Tr\left(\omega^T P^T JP\omega\right) \quad s.t. \quad (P\omega)^T P\omega = L_{n_O} \tag{10.29}$$

The output weight vector if $s > h_n$ is

$$\omega^* = [\tilde{m}_2, \tilde{m}_3, \ldots, \tilde{m}_{n_O+1}] \tag{10.30}$$

Where, normalized eigenvectors.

$$\tilde{m}_i = \tilde{m}_i \,/\, \|P\tilde{m}_i\|, \quad i = 2, 3, \ldots, n_O + 1 \tag{10.31}$$

The output weight vector if $s < h_n$ is

$$\omega^* = P^T [\tilde{b}_2, \tilde{b}_3, \ldots, \tilde{b}_{n_O+1}] \tag{10.32}$$

Where, normalized eigenvectors.

$$\tilde{b}_i = \tilde{b}_i \Big/ \left\|PP^T \tilde{b}_i\right\|, \; i = 2, 3, \ldots, n_O + 1 \tag{10.33}$$

2.2.2.3 RUL performance analysis

Here, multi-ELM-RUL model have been designed based on extracted HI from discharging dataset of battery number 5 and its performance is evaluated in term of absolute (AE), relative error (RE), mean absolute percent error (MAPE), and normalized RMSE as given:

$$AE_{RUL} = |A_{RUL} - P_{RUL}| \tag{10.34}$$

$$RE_{RUL} = |A_{RUL} - P_{RUL}|/A_{RUL} \tag{10.35}$$

$$MAPE_{RUL} = \left(\frac{1}{n} \sum_{i=1}^{n} \frac{|A_{RUL} - P_{RUL}|}{|A_{RUL}|} \right) * 100 \tag{10.36}$$

$$NRMSE_{RUL} = \frac{RMSE_{RUL}}{mean} \tag{10.37}$$

Where, A_{RUL} = actual RUL, P_{RUL} = predicted RUL.

3 Results and discussion

3.1 HI extraction and optimization

3.1.1 HI extraction

As mentioned in Section 2.2.1.1, HI should be easy in measurement and variable with respect to performance of battery. In this study, TI_{Raw} has been evaluated as per Eq. (10.1) and formulated in a series as per Eq. (10.2). The obtained series of TI_{Raw} is plotted for all four batteries (shown in Fig. 10.17) and compared

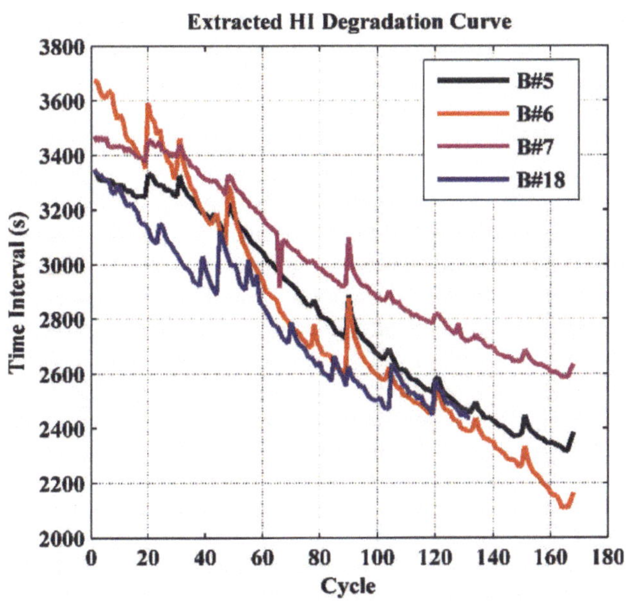

FIGURE 10.17 Extracted HI degradation curve.

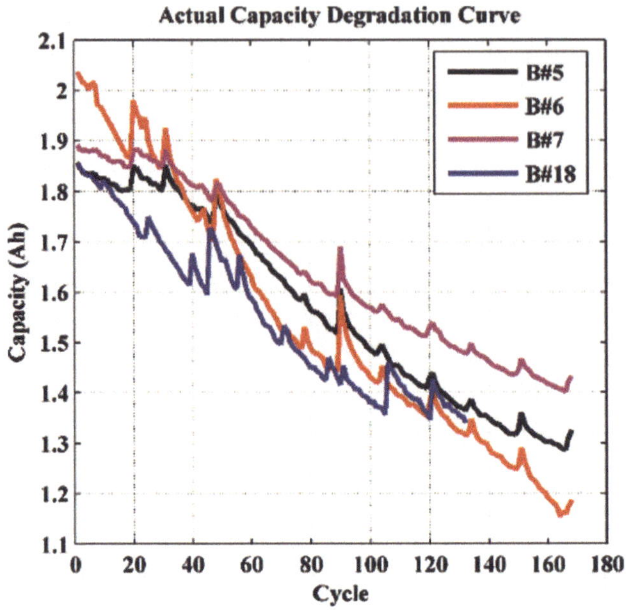

FIGURE 10.18 Actual battery capacity curve.

with actual capacity of the battery (shown in Fig. 10.18). Here, TI_{Raw} is the time interval of discharging condition of the battery.

From Figs. 10.4 and 10.5, it can be concluded that both curve has similarity in pattern. For more validation point of view, a scatter plot has been simulated (shown in Fig. 10.19) between extracted HI and actual capacity of battery, which shows the close linear relationship. Though, there is a need for more improvement in linearity, which can be seen at top portion of the plotted curve.

3.1.2 HI optimization using Box-Cox transformation

In order to improve linearity between extracted TI_{Raw} and actual capacity of battery (Ah), the BC transformation is applied by using R software to search the optimal value of λ. Here, lambda (λ) vary from range +5 to -5 with 1 stepping and a log-likelihood is plotted with 95% confidence interval as shown in Fig. 10.20. From the Fig. 10.20, it is concluded that $\lambda = 1.1248$ has highest performance. Now, a scatter plot is simulated (shown in Fig. 10.21) between optimized/transformed HI and actual capacity of battery, which shows the better linear relationship than Fig. 10.19.

3.1.3 Transformed HI correlation analysis

To analyze the significant improvement in the transformed HI through the BC transformation, Pearson and Spearman rank correlation is evaluated and obtained results are tabulated in Table 10.1. Furthermore, to represent the

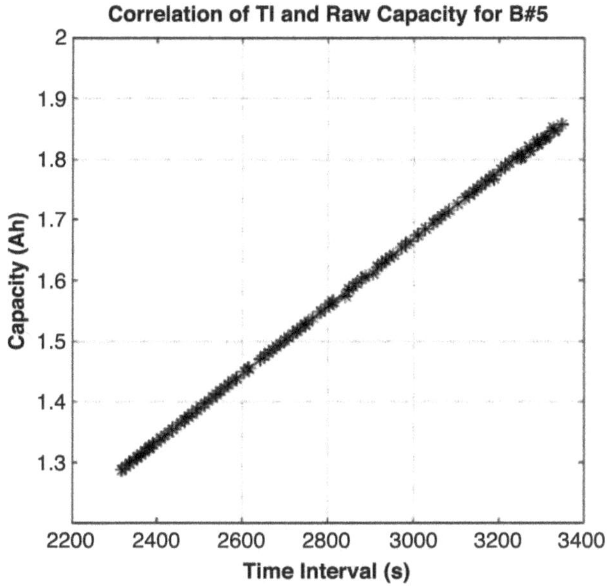

FIGURE 10.19 Scatter plot between extracted HI and actual capacity of battery.

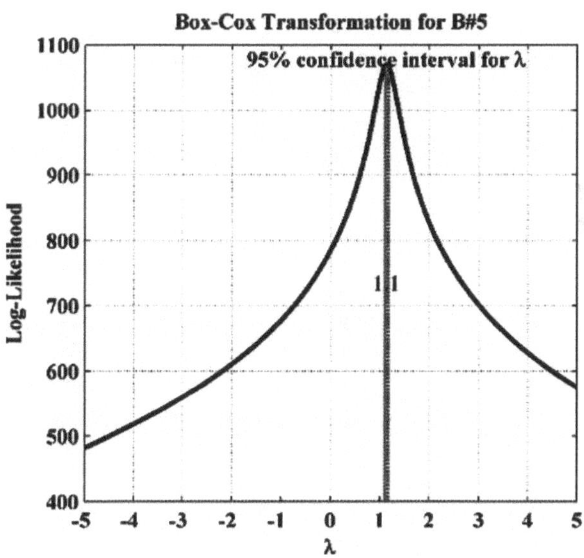

FIGURE 10.20 BC transformation for B#5.

validity of the results shown in Table 10.1, the PCA coefficient is evaluated for all variation of λ from +5 to -5 (same value as used in BC transformation) and results are presented graphically in Fig. 10.22. From Fig. 10.22, it is concluded that PCA coefficient is higher at $\lambda = 1.1248$, which confirmed the reliability of

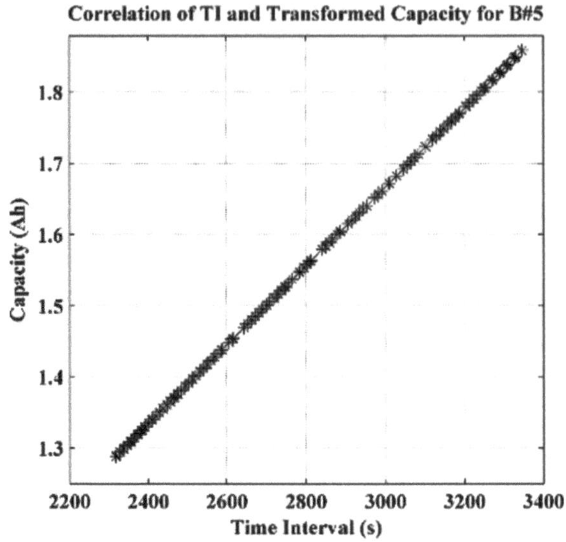

FIGURE 10.21 **Correlation of TI and transformed capacity.**

TABLE 10.1 Quantitative correlation representation for BC transformation.

Time	Pearson correlation	SCA rank	Kendall rank correlation
Before Box-Cox transformation	0.99994489	0.99971907	0.994044
After Box-Cox transformation	0.99998557	0.99982058	0.996738

likelihood method. The highest achieved PCA coefficient is 0.9999 (nearer to 1), which shows a very strong linear relationship between transformed HI and Ah. Moreover, SCA represents the relationship between transformed HI and Ah that HI is canonical enough to replace Ah for identifying the degradation of battery. For more verification, the correlation matrix between HI and Ah is presented in Figs. 10.23 and 10.24 for "before BC transformation" and "after BC transformation," respectively.

3.1.4 HI performance evaluation

After all validation and evaluation as mentioned earlier, the actual capacity and estimated capacity results are presented in Fig. 10.25 for battery number 5–7 and 18. The HI performance evaluation is computed in term of RMSE, R^2, MAPE, and NRMSE and tabulated in Table 10.2 for selected optimal value of lambda (λ) by BC transformation. From Table 10.2, it is concluded that R^2 is very nearer to 1 for all battery cases, which means transformed capacity can be

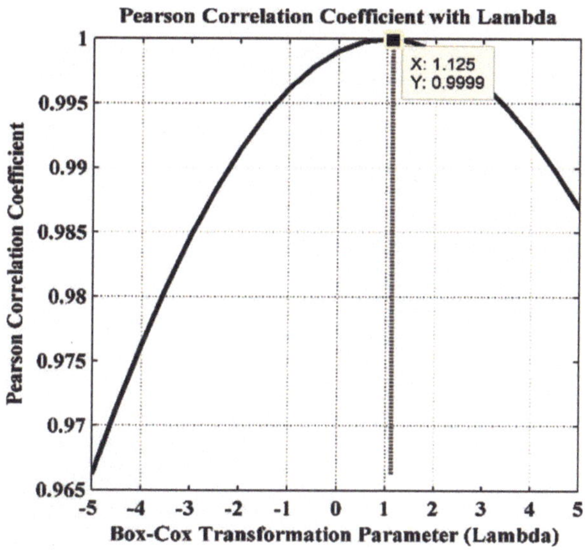

FIGURE 10.22 PCA correlation coeffiecient with λ variation.

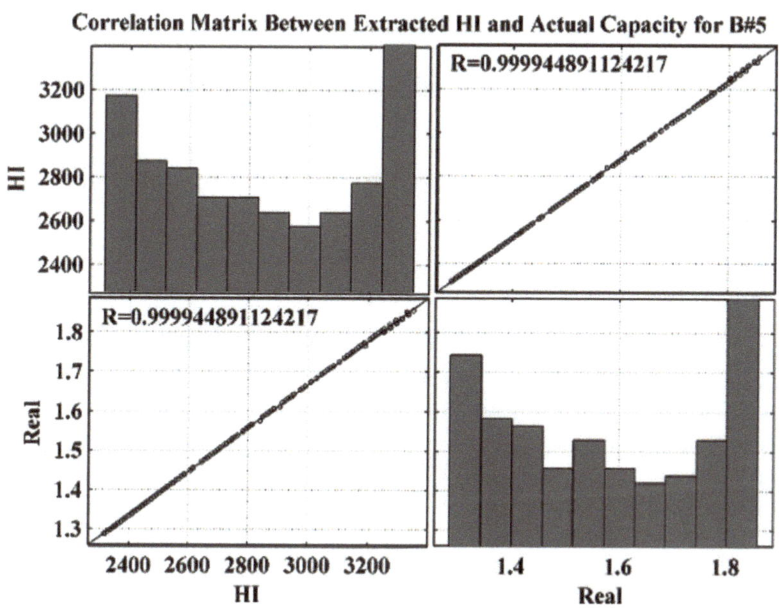

FIGURE 10.23 Correlation matrix between extracted HI and actual capacity of battery #5.

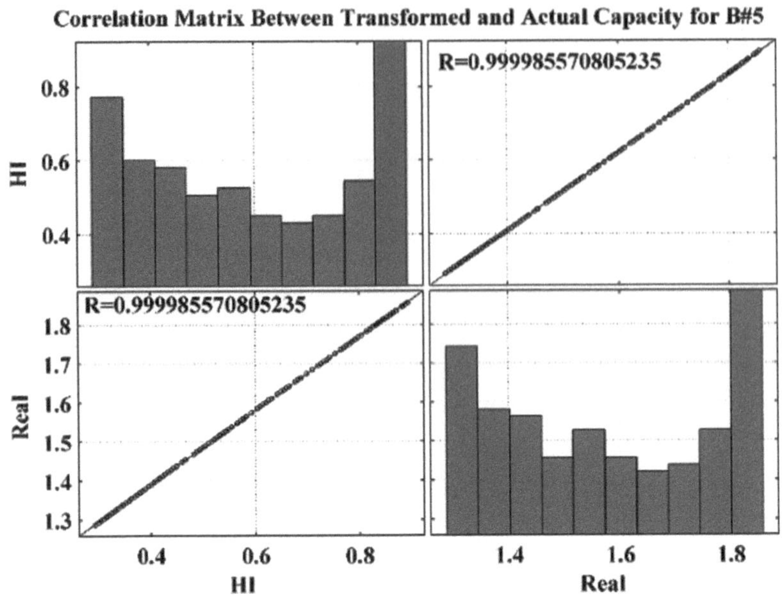

FIGURE 10.24 Correlation matrix between transformed HI and actual capacity of battery #5.

quantify the actual Ah value and hence, real-time degradation can be simulated by using extracted HI.

3.2 RUL estimation using ANN

After analyzing the extracted and normalized HI, RUL evaluation is performed for battery dataset of NASA. The evaluation is carried out into three case studies as follow:

1. ***Case Study#1:*** RUL evaluation from raw HI as input and battery capacity as output to the model. This case study is carried out to validate the correlation of extracted HI with battery capacity.
2. ***Case Study#2:*** RUL evaluation from raw HI as input and transformed HI as output to the model. This study is carried out to validate the consistency of the extracted HI framework.
3. ***Case Study#3:*** RUL evaluation from transformed HI as input and battery capacity as output to the model. This study is carried out to evaluate the similarity of the extracted HI framework in evaluating the RUL.

Now three multi-ELM-RUL models for three case studies have been formulated and tested after proper training & testing phase then RUL has been evaluated for different data samples starting from cycle number 20, 42, 122, 198, and 274 of battery 5.

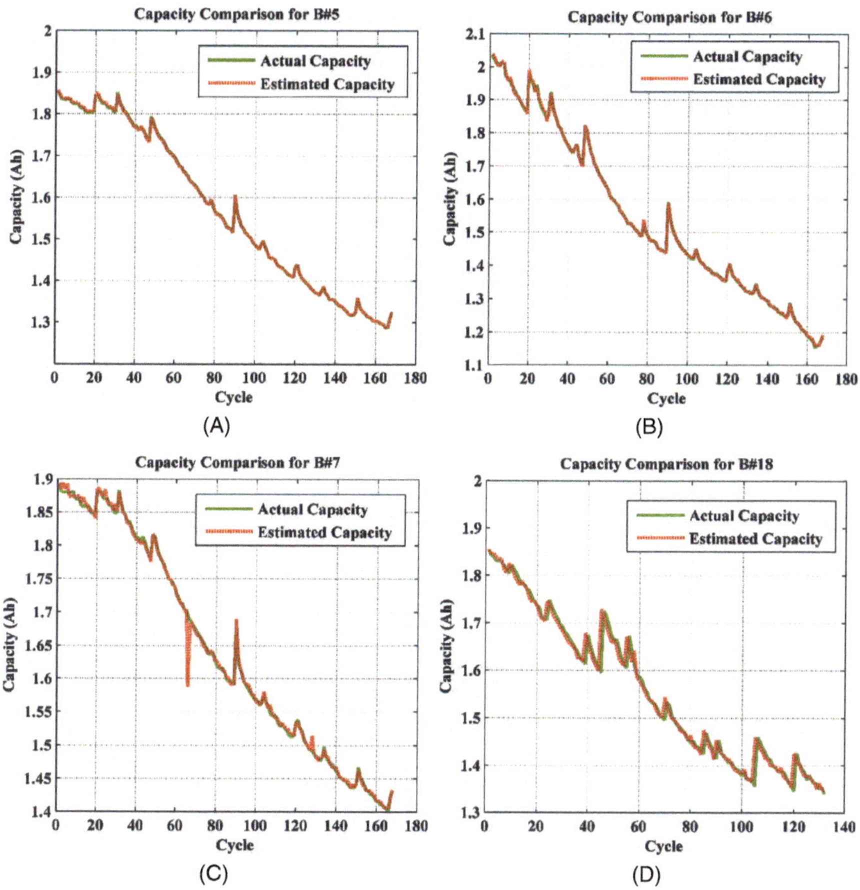

FIGURE 10.25 Estimated capacity and actual capacity comparision representation for (A) Battery 5, (B) Battery 6, (C) Battery 7, and (D) Battery 18.

TABLE 10.2 HI performance evaluation.

Battery	Lambda	RMSE	R2	MAPE	NRMSE
B#5	1.1248	0.001993	0.999998	0.098699	0.003502
B#6	1.0648	0.004331	0.999992	0.219251	0.004913
B#7	1.2378	0.009454	0.999967	0.248994	0.019272
B#18	1.0737	0.022309	0.999796	0.810492	0.043407

Before predicting the RUL, failure threshold of TI is calculated as it is 30% fade capacity rated from 2 to 1.4 Ah for battery number 5. Now BC transformed failure threshold capacity in term of time and cycle number are calculated, which are 2518 s and 124, respectively as shown in Figs. 10.26–10.29.

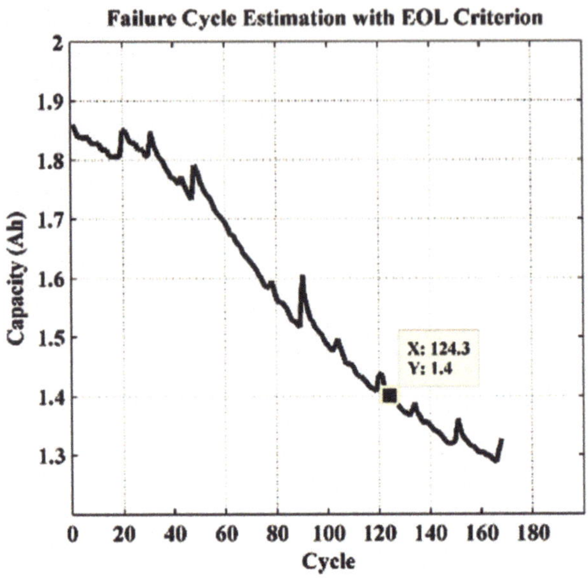

FIGURE 10.26 Failure cycle estimation with EOL criterion.

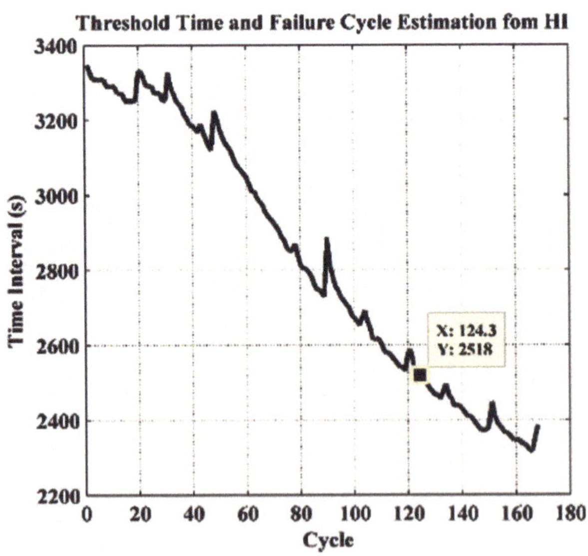

FIGURE 10.27 Threshold time and failure cycle estimation.

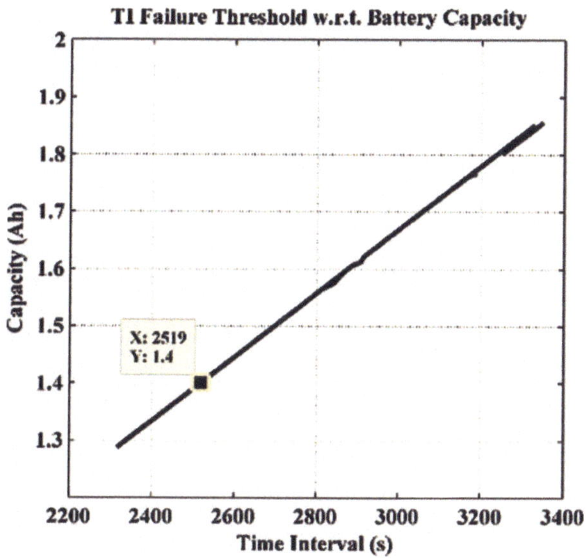

FIGURE 10.28 TI failure threshold with respect to battery capacity.

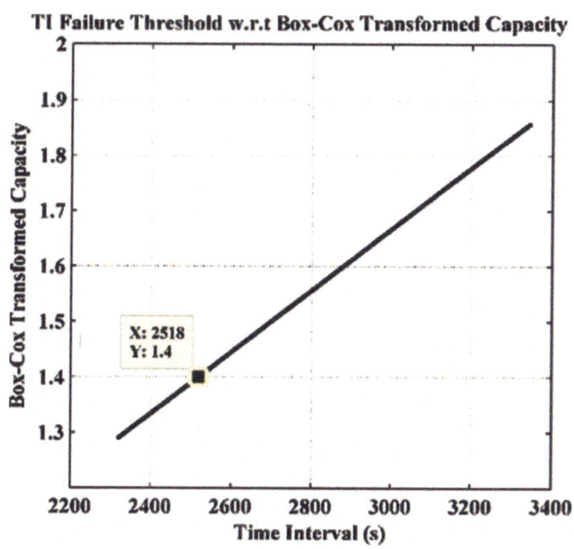

FIGURE 10.29 TI failure threshold with respect to BC transformed HI.

After evaluating threshold value, ANN based RUL estimation is performed and model's performance is evaluated at different dataset starting from cycle number 20, 42, 122, 198, and 274 of battery 5. The RUL forecasting accuracy is evaluated as per Eq. (10.34–10.37) and tabulated in Table 10.3. For better visualization, graphical representation has been shown in Fig. 10.30 for one case.

TABLE 10.3 Case Study#2: SELM-SSELM-USELM-RUL model performance evaluation for B#5.

Starting number	Actual RUL value	Estimated RUL value			Percentage error value (%)		
		SELM	SSELM	USELM	e_{SELM}	e_{SSELM}	e_{USELM}
Cycle number 20	115	113	112	115	1.739	2.609	0.000
Cycle number 42	105	101	100	104	3.810	4.762	0.952
Cycle number 122	85	83	78	84	2.353	8.235	1.176
Cycle number 198	65	61	64	58	6.154	1.539	10.769
Cycle number 274	45	41	43	40	8.888	4.444	11.111

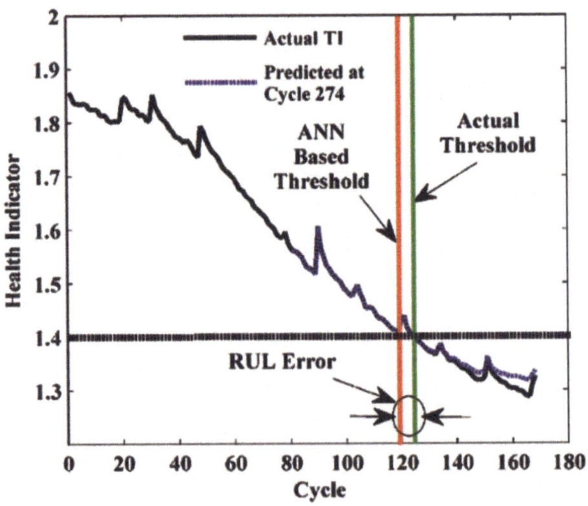

FIGURE 10.30 Estimated RUL representation.

4 Conclusion

In this chapter, a new HI formation and transformation approach for LIB health assessment is proposed. With the help of the power transformation method (named Box-Cox transformation), formulated HI is optimized to increase the linear relationship between extracted HI and the actual capacity of the battery. Thereafter, a linearized relationship is validated by two different approaches, which shows R^2 values near to one. Moreover, multi-ELM methods have been used to realize indirect RUL evaluation with raw extracted, optimized, and transformed HI. Obtained results in each case study satisfy the effectiveness of the proposed approach.

In the future perspective, this approach can be enhanced with other HI variable. Moreover, the selection of the most relevant input variable for RUL assessment needs to be further analyzed.

Acknowledgment

This publication was made possible by Qatar University-Marubeni Concept to Prototype Development Research grant # [M-CTP-CENG-2020-2] from the Qatar University. The statements made herein are solely the responsibility of the authors.

References

[1] https://www.lens.org/(Accessed 14 May 2019).
[2] D. Zhou, L. Xue, Y. Song, J. Chen, On-line remaining useful life prediction of lithium-ion batteries based on the optimized Gray model GM(1,1), Batteries 3 (2017) 1–17, doi: 10.3390/batteries3030021.
[3] N. Williard, W. He, C. Hendricks, M. Pecht, Lessons learned from the 787 Dreamliner issue on lithium-ion battery reliability, Energies 6 (2013) 4682–4695.
[4] Fire, Ambulance and Enforcement Statistics 2017. Singapore Civil Defence Force (SCDF). (2017). Available from: https://www.scdf.gov.sg/docs/default-source/scdf-library/publications/amb-fire-inspection-statistics/fire-ambulance-enforcement-statistics-2017.pdf. Accessed 14.02.2019
[5] Federal Aviation Administration (FAA) Office of Security and Hazardous Materials Safety. Available from: https://www.faa.gov/about/office_org/headquarters_offices/ash/ash_programs/hazmat/aircarrier_info/media/battery_incident_chart.pdf. Accessed 14.02.2018
[6] U.S. Utility-Scale Photovoltaics-Plus-Energy Storage System Costs Benchmark. (2018). Available from: https://www.nrel.gov/docs/fy19osti/71714.pdf. Accessed 14.02.2018
[7] B. Saha, K. Goebel, Battery Data Set, NASA Ames Prognostics Data Repository, NASA Ames, Moffett Field, CA, USA, 2007. Available from: http://ti.arc.nasa.gov/project/prognostic-data-repository.
[8] G. Huang, S. Song, J.N.D. Gupta, C. Wu, Semi-Supervised and Unsupervised Extreme Learning Machines, IEEE Trans. Cybernetics 44 (12) (2014) 2405–2417, doi: 10.1109/TCYB.2014.2307349.
[9] Basic ELM Algorithms. Available from: https://www.ntu.edu.sg/home/egbhuang/elm_codes.html. Accessed 01.08.2020
[10] D. Liu, H. Wang, Y. Peng, W. Xie, H. Liao, Satellite lithium ion battery remaining cycle life prediction with novel indirect health indicator extraction, Energies 6 (8) (2013) 3654–3668.
[11] G.E.P. Box, D.R. Cox, An analysis of transformations, J. R. Stat. Soc. B 26 (2) (1964) 211–252.
[12] W.J. Conover, , Practical Nonparametric Statistics, Wiley, New York, NY, USA, 1999.
[13] D. Liu, J. Zhou, H. Liao, A health indicator extraction and optimization framework for lithium-ion battery degradation modeling and prognostics, IEEE Trans. Syst. Man Cybernetics Syst. 45 (6) (2016) 915–928, doi: 10.1109/TSMC.2015.2389757.

Index

Note: Page numbers followed by "f" indicate figures, "t" indicate tables.

D

E

J

K

L

R

S

T

FSC
www.fsc.org
MIX
Papier | Fördert
gute Waldnutzung
FSC® C083411

Kolibri 360 GmbH
Ferdinand-Jühlke-Straße 7
99095 Erfurt, Deutschland
produktsicherheit@kolibri360.de